Eco-labelling in Fisheries

Eco-labelling in Fisheries

What is it all about?

Edited by
Bruce Phillips, Trevor Ward & Chet Chaffee

Blackwell
Science

Contents

Contributors

Dr Jo AKROYD, Fisheries and Quality Management Advisor, Akroyd Walshe Ltd., PO Box 28–814, Auckland 5, New Zealand.

Dr EDWIN AALDERS, SGS Product & Process Certification, PO Box 200, 3200 AE Spijkenisse, The Netherlands.

Dr ROBERT BOSWORTH, State of Alaska Department of Fish and Game, PO Box 25526, Juneau, Alaska 99802, USA.

Dr LOUIS BOTSFORD, Wildlife, Fish and Conservation Biology, University of California Davis, California 95616, USA.

Dr CHET CHAFFEE, Scientific Certification Systems, 1939 Harrison Street, Suite 400, Oakland, California 94040, USA.

VOLKER KUNTZSCH, Frozen Fish International, GmbH, Am Lunedeich 115, 27572 Bremerhaven, Germany.

Dr DUNCAN LEADBITTER, Regional Director – Asia Pacific, Marine Stewardship Council, PO Box 3051, Stanwell Park, New South Wales 2508, Australia.

BRETT McCALLUM, Executive Officer, Pearl Producers Association, PO Box 55 Mount Hawthorn, Western Australia 6000, Australia.

Dr BRENDAN MAY, Chief Executive, Marine Stewardship Council, 119 Altenburg Gardens, London SW11 1JQ, UK.

Dr PAUL MEDLEY, Sunny View, Main Street, Alne YO61 1RT, UK.

Dr PAUL NICHOLS, No 1, The Waters, Funtley PO17 5EL, UK.

Dr JULIA NOVY-HILDESLEY, Managing Director, The Lemelson Foundation, 721 N.W. Ninth Avenue, Suite 225, Portland, Oregon 97209, USA.

Dr BRUCE PHILLIPS, Department of Environmental Biology, Curtin University of Technology, PO Box U1987, Perth, Western Australia 6845, Australia.

PETER ROGERS, Executive Director, Fisheries WA, Adelaide Terrace, Perth, Western Australia 6000, Australia.

PETER SCOTT, c/o Fulbourne Close, Luton LU4 9PO, UK.

KATHERINE SHORT, Sustainable Fisheries Project Officer, Resource Conservation, WWF-Australia, PO Box 528, 725 George Street, Sydney, New South Wales 2001, Australia.

Dr DAVID SUTTON, Australian Marine Conservation Society (Western Australia), Delhi Street, Perth, Western Australia 6000, Australia.

Dr MIKE SUTTON, Conservation Program, The David and Lucile Packard Foundation, 300 Second Street, Suite 200, Los Altos, California 94022, USA.

Dr TONY SMITH, CSIRO Division of Marine Research, Hobart, Tasmania 7001, Australia.

Dr TREVOR WARD, School of Earth and Geographical Sciences, University of Western Australia, Nedlands, Western Australia 6907, Australia.

Dr MICHAEL WEBER, 228–1/2 South Juanita Avenue, Redondo Beach, California 90277–3438, USA.

Introduction

Bruce Phillips, Chet Chaffee, Trevor Ward & Mike Sutton

1

Eco-labelling is increasingly becoming an important part of the worldwide discussions about natural resources management. *Agenda* 21, adopted at the Earth Summit in 1992, called for the use of eco-labelling to assist consumers to make informed choices (Deere, 1999). This is based on the idea that if consumers are provided with environmental information and a choice between products, many will choose those products that have fewer impacts. Allowing the consumer to select more environmentally friendly products should, in turn, cause producers to alter their behaviour to develop and market products that meet the consumers' requirements (Weber, 2002).

Over the past 20 years there has been an increasing debate about the sustainability of the world's marine fisheries, with many books and scientific papers discussing declining catches worldwide, and providing evidence for the development of a global crisis in marine capture fisheries (see for example Watson & Pauly, 2001). One important aspect of the discussion is the fact that the very agencies that have developed policies to exploit fishery resources are now being tasked with finding solutions. Given the history of fishery policies (see Weber, 2002, as a US example) it is difficult to see how the necessary changes are going to take place. Fishers, conservation organisations, indeed, even government biologists, are expressing increasing concerns over fishing effort, methods of operation and management practices. Yet more and more fisheries continue to join the ranks of those that are being overfished, suggesting that the current tactics of lobbying and litigation are simply not enough to bring about the required changes at a pace fast enough to counteract the decline.

Except for the well-known case of labelling tuna as 'Dolphin Safe', changes in management practices and decreases in ecological impacts have not been significantly linked to eco-labelling practices. In fact, there is still little worldwide data on the consumer product-based eco-labelling programmes (more than 40 programmes; EPA, 1998) to show that any significant environmental improvements have been achieved through eco-labelling of products. Nevertheless, in the past 10 years there has been a surge of interest and activity in seafood eco-labelling. National and international government agencies, industry organisations, and environmental groups have all launched schemes, including seafood guides, about sustainably produced seafood (Deere, 1999; Duchene, 2001).

At the 2nd World Fisheries Conference in Brisbane 1996, Australia, Sutton (1997) presented a paper titled 'A new paradigm for managing marine fisheries in the next millennium', which characterised the state of the world's fisheries, and pointed out why current strategies by conservation groups were falling short of securing the reforms necessary to manage fishery resources sustainably. In the paper, Sutton identified the fact that the typical tools – lobbying, campaigning, and litigation – were useful but were geared toward fighting industry and government, so by their very nature were divisive and slow to produce change. Taking stock of other efforts around the globe where conservationists were actually entering into partnerships with industry to effect change in natural resource management, Sutton proposed that the same practices could take place in fisheries. The idea was to create a new tool by introducing an eco-labelling system, to identify 'well-managed' fisheries, thereby offering a reward to industry for good management.

The following year, 1997, saw the establishment of the Marine Stewardship Council (MSC) in the UK – an organisation founded through a partnership between WWF, the world's largest conservation organisation, and Unilever, one of the world's leading seafood buyers. By December 2001, six fishery projects, covering tens of fisheries, had assessments completed leading to certifications by the MSC. These projects led to eco-labelled seafood products from certified fisheries available for sale in the UK, Switzerland, Belgium, Germany, France, The Netherlands, Australia, and the US.

Despite all the discussions, workshops, trade shows, and conference presentations on the subject over the past 6 years, eco-labelling in fisheries is still not well understood by most fishery managers, scientists, members of the fishing industry, or conservation organisations. Some have a vague idea of the concept, but no in-depth knowledge about how assessments are conducted, what technical expertise is necessary, how certification programmes can be used to modify management practices, or the ways in which eco-labels might help both industry and conservation groups to achieve their objectives.

As the next generation of fishery biologists and resource management professionals are trained, exposure to this subject will be increasingly important if these programmes are to grow and adopt better practices to gain maximum benefit. At present, three Australian universities have course components covering the concepts of eco-labelling in fisheries, and three others – Stanford University Business School, USA; IMD, Switzerland; and Oxford University, UK – have sponsored case studies on the business application and benefits of programmes such as that of the MSC.

This book has been written to help provide some basic information about the current process of eco-labelling, and the possible outcomes that it may produce in wild capture fisheries. All of the authors contributing to this book have had experiences of eco-labelling, in one way or another. Their experiences are not uniform, and they have seen considerable variations in the processes necessary to undertake the assessments of different fisheries.

The MSC features prominently. Although this book is not meant to be exclusively about the MSC, it is the most comprehensive, and the only global scheme, that really addresses ecological issues. We have encouraged contributions to this book from all perspectives, seeking at all times to make any criticism of MSC or eco-labelling as constructive as possible. We offered people with a range of interests and experience an opportunity to contribute, and although not all were able to write a chapter, the material published here represents a reasonable spread of experience and opinion about eco-labelling and the MSC programme.

There have been many articles in the press discussing sustainability and eco-labelling. Some criticise the MSC processes or the lack of applicability to all fisheries (Viswanathan & Gardiner, 2000; Dunlop, 2000; Warren & Haig-Brown, 2002). Some report increased profitability for their company after certification (Hedlund, 2001), while others simply discuss the concept (Wessells, 2000; Pizzico, 2002). However, all of this increases the need for broadly-based and accurate information about the principles and practice of eco-labelling in fisheries, so that students, scientists, resource managers and seafood consumers can be properly informed.

Eco-labelling in fisheries is rapidly changing on a day-by-day basis and we are sure that a revision of this book will be necessary in only a few years time. The MSC is also rapidly changing itself, so, any specific aspects discussed in this book, such as certification procedures, dispute resolution procedures etc, should be checked against the MSC website (www.msc.org).

Seafood Evaluation, Certification and Consumer Information

2

Chet Chaffee, Duncan Leadbitter & Edwin Aalders

2.1 Introduction

Certification is not a new idea, either generally or as it relates to eco-labelling in fisheries. Certification is simply the act of assessing and verifying information. In most instances, this involves comparing a fishery to some stated standard (either self-determined or independent). Under the International Standards Organization (ISO) this is often referred to as 'conformity assessment', however, depending on the country, the assessment of information for compliance with a known standard can also be known as registration or certification. For example, in the United States auditors assessing companies for compliance with ISO 9000 (Quality Assurance) or ISO 14000 (Environmental Management) standards use the term registration as certification legally implies added aspects of liability.

Certification has been around for hundreds of years providing assurances for almost everything we use from the services of professionals to the food we eat. There are literally hundreds of certification programmes run by national, regional, and local governments as well as by non-governmental and professional organisations. For example, there are programmes in almost every country to certify human capabilities such as drivers of cars, pilots of ships, auditors of banks, policemen, firemen, doctors, nurses, dentists, engineers, laboratory technicians and just about every other professional position. In addition, we certify all types of products from the design plans for buildings, to the quality of products we purchase to the financial statements of publicly traded companies.

Certification for the purposes of product labelling is also not new. In 1894, underwriter laboratories began certifying and labelling the safety of products in the United States. The labelling of environmental aspects of products has a history that dates back to at least 1977 with the inception of the Blue Angel environmental labelling programme in Germany. The globalisation of businesses coupled with increased consumer access to information caused a proliferation in environmental certification and labelling programmes from the late 1980s. In 1992, 20+ environmental labelling programmes had been developed worldwide (EPA, 1993). By 1998, more than 50 environmental labelling programmes could be identified as in effect or developing (EPA, 1998). The environmental labelling programmes reviewed by the US Environmental Protection Agency (EPA) are only the tip of the iceberg with regard to environmental certification, as they do not include food certification and labelling, which includes organic labelling, biological contamination or food safety, and genetic modification.

2.2   Certification for seafood

One of the primary functions of certification as regards the dissemination of environmental information is that it assures consumers that they are receiving verifiable, accurate, non-deceptive information (ISEAL, 2001). Deere (1999) points out that the credibility of an eco-labelling programme greatly depends on the technical integrity of the certification process. Certification can be equally as important in non-labelling programmes, such as ISO 14000, as it still provides the guarantee that the service or product being assessed complies with a relevant standard.

The technical integrity of any certification process is dependent on many things; however, the primary concerns for maintaining information integrity and consumer confidence are the development of standards, the methods of assessment, and the integrity and capability of the assessors.

2.2.1   Standards

Weber (2002) provides a summary of the types of standards commonly found associated with the dissemination of environmental information or product claims and the barriers that standards can evoke. It is generally accepted that there are three types of standards:

- First-party standards – developed by a company for application within the company,
- Second-party standards – developed by an industry organisation to apply to the entire industry sector,
- Third-party standards – developed by organisations independent of the industry to which the standards apply.

Any of these types of standards can be applied to seafood depending on the objective of the company or organisation utilising them.

First-party standards are typically used by the company that develops them. However, they are open to significant criticism concerning conflict of interest.

Second-party standards are often seen as more credible than first-party standards (Weber, 2002) and are more widely implemented. However, these standards too are open to criticisms. In the case of second-party standards the common criticisms are conflict of interest and playing to the lowest common denominator. Industry associations are usually in the position of having to protect all members equally, so creating standards that allow for significant differentiation among members is often against the charter of the association, if not the express wishes of its paying members.

Third-party standards are often seen as the preferred choice in supporting the dissemination of environmental information to consumers as they have the appearance of independence and objectivity. However, the true objectivity of these standards is dependent on the range of input provided during the standards development process. As Weber (2002) points out, the preferred method of standards development to maintain true objectivity and independence is to include a wide range of stakeholders. Without a wide range of stakeholder input, third-party standards can become as biased as any other standard. Where third-party standards have been developed with little or no industry representation, industries, including seafood, have criticised third-party environmental standards as being heavily biased against them.

In addition to the types of standards, Deere (1999) notes that the geographic origin and scope of a standard can also cause problems. National or regional standards can be used as trade barriers for producers outside the region. Also, standards developed in industrial countries may prove difficult for developing countries to comply with due to costs of certification, lack of technology, or costs associated with bringing operations into compliance with outside standards.

2.2.2   *Methods of assessment*

Another important aspect of certification is the method utilised to evaluate the environmental aspects of a fishery. Assessment methods comprise two important variables: (a) the evaluation process and (b) the information evaluated. Broadly speaking, the more rigorous the methods used to assess the environmental aspects of a product or process and the more depth and breadth in the sources of information, the more credible the system.

The evaluation process can range from identifying the necessary components in good fisheries management to evaluating the performance of each component. In the first instance, an assessor has a checklist of the components required for good management and simply seeks evidence that these components are in place and utilised. Many believe that if the standard used is specific enough, the mere identification of components can be a reasonable proxy for identifying good management. Detractors are much less forgiving, commonly stating the concern that procedures may be in place but not properly administered. For example, a stock assessment may be required in a fishery but the assessment may utilise an improper model, bad assumptions, and faulty tests of reliability. A certification is often considered more credible if the evaluation process not only identifies the necessary components but also has some means of evaluating their performance.

Information content is also important during certification. The information required by certification programmes ranges from signed affidavits to data gathered from on-site audits. As the breadth and depth of information increases, so does the credibility of the programme. For example, affidavits from companies, even when signed by an officer of the company, are fraught with problems of conflict of interest. The stated integrity of having signed affidavits from company officers is that false submissions have negative legal ramifications. Without significant negative legal consequences or the ability to check the information, affidavits can be suspect. Desktop reviews of information submitted by a fishery provide a slightly better means of assessment; however, these too are subject to bias. To improve the situation, desktop reviews can be performed using additional information gathered independently by the assessor from publicly available sources that have been peer reviewed. To further improve the breadth of information in an assessment, interviews with key personnel in the fishery can be conducted. In interviews, assessors can inquire how management decisions are made, how research agendas are set, what assessment models are used, and many other aspects of a fishery where more information is needed. Adding audits to the evaluation process provides yet another level of assurance. Audits allow the assessors to review documents on-site in management agencies, research organisations, and fishing companies to verify information received. Last, assessors can solicit information from stakeholders in a fishery. In many instances, this represents the only opportunity for assessors to hear about unintended consequences of fishery management, thus providing some balance to the often favourable information found in the literature or provided by the fishery.

2.2.3   Qualifications

Another important aspect of certification is who is conducting the assessment. Certifications range from self-certification to third-party certification. Self-certification, as its name implies, is conducted by personnel within the fishery. Like first-party standards, the potential conflict of interest here can be troubling, even if the assessors use the most objective standards and the best information. Third-party assessments at the very least provide the air of independence and objectivity. However, the full potential of third-party audits are only achieved if the stakeholders in the fishery and the consumer believe that the assessors were properly qualified and objective. Programmes that include an accreditation system for qualifying assessors and an assessor selection process that utilises input from a wide variety of stakeholders provide the most integrity and transparency.

2.3   Environmental information for seafood

There are several mechanisms used to communicate the environmental aspects of seafood to consumers. They range from product labels, to information brochures and cards identifying seafood choices, to marketing and advertising materials.

Casio (1996) describes nine principles at the heart of the ISO standards for those developing and using eco-labels and declarations, most of which would appear to be appropriate for any system providing information to the public:

(1)   Labels and declarations must be accurate, verifiable, relevant, and non-deceptive.
(2)   Environmental attributes must be available to purchasers.
(3)   Labels and declarations must be based on thorough scientific methodology.
(4)   Criteria for label or declaration must be available to interested parties.
(5)   Label and declaration development must take into account the life cycle of the relevant product or service.
(6)   The administrative work must be limited to establishing compliance with criteria.
(7)   Label and declaration procedures and criteria must not create unfair trade restrictions or discriminate between foreign and domestic products or services.
(8)   Labels and declarations must not inhibit innovation.
(9)   Label and declaration standards and criteria must be developed by consensus.

The embodiment of these principles goes to the heart of any seafood eco-label as well. Specifically, principles (1), (2), (7), and (9) point to the same criteria discussed throughout this chapter for developing a credible certification programme.

Two types of labels are typically found on seafood products: seal-of-approval labels, and single attribute labels. Both of these label types have been reviewed by ISO with guidance provided on how to set up and apply programmes properly for these types of labelling in the 14000 standard series.

Single attribute labels, by definition, refer to a claim on a product that is in reference to one attribute. 'Dolphin Safe' labels on canned tuna refer to a single attribute, as does 'Turtle Safe' shrimp. Although not a seafood product, 'Salmon Safe' wine is also a single attribute label espousing the benefits the product brings to widely known and valued fish species. While explicit, the standard complaint about these types of labels is that they do not take a broad enough view of the environmental implications associated with the product. While many tuna fisheries may now have solved the problem with dolphin by-catch, there are many that are facing problems of reduced stock size, by-catch of other fish species, and lack of agreed management measures.

Seal-of-approval labels are based on the product meeting a pre-determined set of attributes that typically cover a broad range of issues. In many seal-of-approval programmes, the entire life cycle of a product is considered. For seafood, this might well mean taking into consideration all aspects of the managed fishery from the early life history of the targeted species to the strategies used to control the harvest of the species. However, the set of pre-determined attributes may comprise only a part of a fishery as the developers of those systems have the freedom and flexibility to define the boundaries of the system. The Marine Stewardship Council certification pro-gramme is an example of a programme that provides a seal of approval for a fishery upon the successful determination by a certifier that the fishery is in compliance with the MSC standards, which cover most aspects of fishery management from basic research through management and compliance. Both single attribute labels and seal-of-approval labels are positive declarations about seafood that has been shown to be in compliance with a stated standard. Neither type of programme explicitly denigrates any other fishery or seafood product.

Guides to sustainable fisheries or sustainable seafood have surfaced over the past few years. While not certification programmes as described above, they are another form of disseminating environmental information. The reason for including them in this discussion is that they are growing in occurrence and popularity, and are mostly produced by conservation organisations seeking to provide members and the general consumer with a different view of fisheries management than other systems. For example, these systems often put out consumer friendly cards or brochures showing what seafood has been determined to be good, and in contrast to the product labelling approaches outlined above, include negative information on what products to avoid or to purchase with caution. To support the advice provided, these systems often publish a report showing how conclusions were reached and the methods used. It is not surprising to find that there are opponents to this type of system as well. The opponents to the current suite of rating systems or guides have expressed a number of concerns stating that these programmes are seriously flawed because they do not have comprehensive criteria, the evaluation process is too subjective, they lack the ability to tackle problems with technical rigor, they rely on limited information, and they are not fully transparent.

It is interesting to note the trend in opponents to different systems of information dissemination. Opponents to eco-labels in the environmental and conservation organisations are sceptical about the seafood industry's interest in getting certified. Many have expressed concern that certification programmes are not rigorous enough in terms of defining sustainable management and that they have not incorporated enough participation by environmental groups. In contrast, rating systems appear to be mostly opposed by the seafood industry. Part of the opposition appears to be based on the fact that these programmes are almost solely run by conservation organisations that also spend a good deal of their time and money fighting fisheries management agencies. In addition, the industry has expressed concern that these programmes provide negative ratings about fisheries through a proprietary evaluation process that is not open to participation by all stakeholders.

2.4 Other certification programmes

Marketing materials, brochures, and advertisements can also be used to disseminate information about the environmental attributes of seafood. These pieces can state as much or as little information as needed to convey a message as long as it is not deceptive. In some cases this form of disseminating information is used because product labelling is not allowed (e.g. ISO certification such as ISO 14000 or ISO 9000), in other cases it may simply be a matter of choice.

HACCP (Hazard Assessment Critical Control Point), a food safety programme that controls for biological and chemical contaminants, is required in a number of European countries, the United States, Canada, Australia and New Zealand. Some HACCP certification programmes do not allow labelling such as those offered by government agencies or those run under ISO *Guide 62 Accreditation*, while some third-party companies not only offer HACCP but offer a label as well (e.g. Scientific Certification Systems' *CertiClean* programme in the United States). While these types of certification are offered for seafood, they are very different from the seal-of-approval and single-attribute programmes noted above in that they do not deal with the issues of sustainability of target resources or impacts on the marine environment. As a result, they are not dealt with any further in this chapter.

ISO 9000 and 14000 do deal with management systems; however, ISO 9000 focuses only on quality systems. ISO 14000 does focus on environmental systems, but different from eco-labelling programmes, it does not require that fisheries meet independently derived standards. Under ISO system certification, the overall setting of objectives (also known as standards) is determined internally by the entity seeking ISO certification. The business entity seeking ISO certification or registration needs to state its objectives and show continuous improvement over time in meeting these objectives. In this way it is very different from programmes that only grant certifications where entities (fisheries in the case of ISO 14000) can prove they meet an independent standard developed through a consensus approach involving a wide variety of stakeholders. This difference has been cited by major environmental and conservation organisations as a serious concern, since it means that industry can self-determine the level of sustainability to which it wishes to aspire. After three years of certification, a certified company could still be the worst within its own industry sector yet maintain its certification if it has shown continuous improvement against its self-defined objectives. It is not surprising, therefore, to find that conservation groups have complained that fisheries will simply use the ISO process as a device to fool consumers, without meeting the level of management that many stakeholders in the conservation sector believe is necessary to provide truly sustainable fisheries. We believe that if a fishery seeking ISO certification used an outside standard, such as that drafted by the MSC or the guidelines drafted by the UN Food and Agriculture Organisation (FAO), as its set of objectives, conservation groups might find this more appealing and supportable.

Another programme category is government-mandated programmes such as the guidelines for gaining export approval under the *Environment Protection and Biodiversity Conservation Act*, 1999, in Australia (Schedule 4 listing under the *Wildlife Protection (Regulation of Exports and Imports) Act* 1982). This programme requires that all fisheries in Australia provide proof that they are meeting a standard for ecologically sustainable development (ESD) before processors can export product obtained from that fishery. This programme, different from ISO 14000, covers issues of resource sustainability and environmental impacts using standards developed through the Australian government processes. In this programme fisheries are allowed to self-report (similar to self-certifying the information), and the government agency tasked with issuing the certificate (Environment Australia) must review and approve the information. Sceptics have suggested that the government review is too lax, and will not hold fisheries in Australia to tough enough standards in terms of data verification and review.

2.5   Sustainable seafood programmes – an overview

A few labelling programmes in fisheries have been around for more than 20 years. More recently, quite a number have sprung up as conservation groups and industry alike look for new ways to achieve and promote more sustainable fisheries management through market-based incentive programmes. Studies from a variety of industry sectors have shown that attaching claims of environmental preference can have a positive effect on purchasing behaviour, and the same is expected for seafood. The underlying premise is that putting environmental information in the hands of consumers can help promote the purchase of these products, which in turn provides economic benefit to those fisheries providing environmentally beneficial management (Wessells *et al.*, 2000). More importantly, those interested in supporting improved conservation measures in fisheries management realise that economic incentives can play a positive role in achieving those goals if they are used judiciously and in conjunction with other proven methods such as litigation and campaigning.

Some examples of programmes that provide ecological or environmental information on seafood are shown in Table 2.1. The programmes identified were chosen for illustrative purposes only and are not intended as a comprehensive list of evaluation, certification, or government-permitting programmes. Rather, the examples provided are to illustrate the differences in types, quality, and verifiability of information used in programmes asserting that they evaluate the environmental consequences of extracting seafood from the marine environment. There are many other programmes that have not been mentioned in both capture fisheries and aquaculture, including many company-specific buying specification programmes that are used to evaluate corporate supply chains. However, these types of programmes are not often used to make any public claims regarding sustainability to the general public either through government permitting, eco-labelling, or direct publication of consumer information.

The programmes listed in Table 2.1 vary in that some focus on capture of wild fish, others on farmed or aquaculture products, and still others on both. Some produce only positive information, while others also provide negative information. The characteristics of these programmes are outlined to illustrate the differences. As we have pointed out, the most comprehensive programme would be one that develops standards independent of the fishery or of the seafood industry and uses a consultative and consensus process that includes all stakeholders including industry and conservation groups, requires a performance evaluation that includes audits and input from stakeholder groups outside the fishery, and uses external reviewers (also known as auditors or evaluators) that are not part of the staff from the programme initiative or the fishery. Two characteristics that are not identified in Table 2.1 are the use of peer reviews to provide extra assurances that the data and conclusions are accurate and well substantiated, and transparency. Few programmes appear to require stepper review; however, it has been noted as an appropriate step under ISO and is known to occur in the Marine Stewardship Council programme. Transparency is a goal that many of the programmes cite but is not fully applied to enable any interested party to obtain information about the program, its employed methods, and the resulting conclusions.

The Marine Stewardship Council programme was developed to contain many of the best characteristics, including peer review and transparency, and in so doing provides the best choice in seafood labelling and information programmes. The

Table 2.1 Overview showing examples of types of evaluation, certification, and labelling programmes for seafood.

Programme	Organisation or Agency	Type of Programme				Standards				Assessors			Assessment Methods			
		****Vol.	***Req.	Wild Fish	Farmed Fish	Single Attribute	Multiple Attribute	Internal	External	Internal to Fishery or Industry Sector	External to Fishery or Industry Sector **Programme Dependent	*Programme Independent	Check List	Performance Evaluation	Audits	Stakeholder input
Single Claims Programmes																
Dolphin Safe Tuna	US Govt. & Earth Island Institute		×	×		×			×						×	
Turtle Safe Shrimp	Earth Island Institute	×		×		×			×				×		×	
Government Programmes																
Ecologically Sustainable Development	Australian Federal Government	×	×	×	×		×		×	×				×		
Eco-label Programmes																
Sustainable Fisheries Management	Marine Stewardship Council	×		×			×		×			×		×	×	×
	Fundación Chile	×		×	×		×	×		×			×		×	
Sustainable Production and Organic Processing	Farm Verified Organic (Independent company)	×		×			×		×		×		×		?	
Responsible Aquaculture Programme	Global Aquaculture Alliance	×			×		×	×				?	×		×	
EMS Certification Programmes																
14001	ISO	×		×	×		×	×				×	×		×	
Rating Systems or Guides																
Seafood Lover's Guide	National Audubon Society	×		×	×		×		×		×			×		?
Seafood Choices	Monterey Bay Aquarium	×		×	×	×	×		×		×			×		?

**** Voluntary
*** Required
** Programme Dependent – means the reviewers are part of the staff in the programme, rather than independent reviewers.
* Programme Independent – means the reviewers are not programme staff, they are hired as external reviewers.
? Means the issue is in question either because it has not been decided or it has not been implemented.

questions on many people's minds are how the programme works and whether it is achieving its goals. Further chapters in this book review the various steps in the MSC programme by a number of authors that have participated either on evaluation teams, as peer reviewers, or as stakeholders. Their views provide some insight into how the MSC programme functions, and whether it is working.

The Marine Stewardship Council (MSC) Background, Rationale and Challenges

Brendan May, Duncan Leadbitter, Mike Sutton & Michael Weber

3

3.1 Introduction

The world's supply of seafood comes from four basic sources: marine and freshwater capture fisheries, and marine and freshwater aquaculture. Keeping in mind that global statistics on fisheries' production should be treated with caution, global fish production was 126 million tonnes (Mt) in 1999, according to the UN Food and Agriculture Organization (NMFS, 2000). While landings from capture fisheries have levelled off generally and declined in many cases, aquaculture production has increased steadily so that it accounted for most of the growth in fisheries production in the 1990s. In 1999, marine capture fisheries produced an estimated 84.6 Mt, while 8 Mt came from freshwater capture fisheries. World aquaculture production rose from 13 Mt in 1990 to 33 Mt in 1999. Of 1999 aquaculture production, 20 Mt was from freshwater and 13 Mt from marine.

Hundreds of species make up world catches from marine fisheries; however, several groups of species dominate in volume, while others dominate in value. High volume species groups include sardines-anchovies-herrings, cods-hakes-haddocks, tunas, jacks, mackerels, and salmon (NMFS, 2000). While some of these species groups, such as cod and salmon, are staples of seafood consumption in industrialised countries, other species groups such as sardines are destined for conversion into fish meal and oil. High-value but low-volume species groups include shrimp and lobsters. Worldwide, aquaculture now produces 60% of the world's salmon and 28% of the world's shrimp.

Official statistics, which can be quite unreliable depending upon the country, show that China leads all other countries in production from both aquaculture and capture fisheries by a wide margin (NMFS, 2000). Other countries leading in capture fisheries, in order, are Peru, Japan, India, Chile, the United States, and Indonesia. (Peru's landings are dominated by anchovies, which fluctuate widely and do not enter markets for direct human consumption.) Other countries leading in aquaculture production are India, Japan, Indonesia, Thailand, and Bangladesh.

Increasingly, fish are used for direct human consumption rather than for conversion into fish meal, which in turn is fed to cattle, poultry, and fish. Nevertheless, in 1999, an estimated 21% of fish production was reduced to fish meal and oil, but of the remainder 35% was marketed fresh, 21% frozen, 11% canned, and 9% cured (NMFS, 2000).

The chain of custody for seafood products is extraordinarily diverse in type, and often quite lengthy and complex. Producers range from very small-scale, artisanal, local fishers in both developing and industrialised countries to very large-scale, industrialised, distant-water fishermen. Processors and distributors are similarly diverse in size and distance from the consumer, ranging from fishermen who sell fresh fish directly to the public to large, industrial scale, vertically integrated processing and distributing companies that sell into national and international markets. Markets range from dockside sales to supermarkets, restaurant chains, and large institutions such as schools.

Roughly one third of all fish production enters international trade (FAO, 2000). In 1998, international trade in fishery commodities (both for human consumption and for industrial purposes including fish meal) generated an estimated US$55 billion in imports (NMFS, 2000). Most exports are frozen, chilled, or fresh fish. For quite some time, there has been a net flow of fish from developing countries to industrialised countries, especially the United States, the European Union, and Japan. Developing countries accounted for 20% of exports by volume and nearly 50% by value in 1998 (FAO, 2000). In 1998, developing countries registered a trade surplus of $17 billion. Of the $15 billion of exports in 1998, the top five countries of Thailand, Norway, China, the United States, and Denmark accounted for one third of the total (NMFS, 2000). As in previous years, Japan led all other countries in fish imports, accounting for 23% of the total, followed by the United States with 16%, and member states of the European Union accounted for roughly 30% of all imports.

3.2 The problem

Although different organisations state the problem of marine fisheries somewhat differently, they commonly identify a number of broad problems that have led to declines in fisheries and to disruption of aquatic ecosystems. These problems include:

- overfishing fostered by inadequate regulation and oversized fishing fleets;
- wasteful fishing practices;
- damage to habitat caused by certain fishing practices and by coastal development and pollution.

In initiating the process that led to the formation of the Marine Stewardship Council in 1997, the WWF (formerly the World Wide Fund for Nature) further concluded that continuing to rely on government management programmes alone, without engaging the consumer, would fail to stem the decline of fisheries and marine eco-systems.

3.3   The theory of eco-labelling

While other schemes for eco-labelling seafood have been developed, the most comprehensive and ambitious is the Marine Stewardship Council (MSC). The MSC is the focus of this section, although other initiatives are described and discussed at the end.

The foundation of the initiative by WWF and Unilever PLC, a multinational consumer goods conglomerate, was the theory that harnessing consumer purchasing power could generate change and 'promote environmentally responsible stewardship of the world's most important renewable food source'. The MSC claims that its eco-label will translate complex scientific information into a simple message that can be understood by the consumer so that the consumer can purchase seafood that has minimal impact on the aquatic environment. These purchases in turn will provide an incentive for fishing responsibly. According to the MSC, fish processors, traders, and retailers who buy from certified, sustainable sources will benefit from the assurance of continuity of future supply (MSC, 2001a).

The MSC believes that the fishing industry will benefit from MSC certification in the following ways (MSC, 2001a):

- evidence and recognition of good fishery management;
- improved management of fisheries;
- preferred supplier status;
- potential for improved returns;
- new markets.

A 1999 survey of US consumers sponsored by Rhode Island SeaGrant argued that the 'critical factor that will determine the success or failure of eco-labels is consumer acceptance' (Wessells, *et al.* 2000).[1]

3.4   Brief history of the MSC

The MSC was founded at the initiative of the WWF and Unilever, one of the world's largest buyers of frozen fish, in 1997 (MSC, 2001b). The MSC grew out of a business-environment partnership that the two organisations, both leaders in their respective fields, formed in 1995. Although the motivations of the partners were quite different,

---

[1] The study relied on a survey of 1640 seafood consumers across the contiguous United States in September and October 1998. The survey focused on salmon, cod, and shrimp, because they are among the most popular seafood products in the United States.

their goal was the same: to reverse the increasingly serious trend of unsustainable fishing. WWF was concerned about the widespread impact of overfishing on marine ecosystems and the limited capacity of regulatory programmes to ensure that fisheries are sustainable. Unilever argued that the future of some of its brand name companies that deal in frozen seafood, such as *Birds Eye* and *Iglo*, was threatened by growing consumer perceptions about the oceans and potential future interruptions of supply caused by overfishing. The two partners decided to create the MSC to harness market forces and consumer power in favour of sustainable, well-managed fisheries. Together, they contributed more than US$1 million in start-up funds to the MSC, evenly split between the partners. Since 1999, the MSC has operated independently of its founders.

The MSC states its mission as 'to safeguard the world's seafood supply by promoting the best environmental choice'. Within this mission, the MSC says that its duties are:

- to conserve marine fish populations and the ocean environment on which they depend;
- to conserve the world's seafood supply for the future;
- to provide consumers with accurate information about the best environmental choice in seafood;
- to work in partnership with its stakeholders;
- to ensure its programme and its benefits are available to all regardless of size or region;
- to carry out its activities responsibly and openly.

The MSC says that it will do the following, among other things, in carrying out this mission:

- encourage independent certification of fisheries to the MSC Standard;
- identify, through the MSC eco-label, products from certified fisheries;
- encourage buyers and sellers of seafood to source MSC-labelled products;
- assess and accredit independent third-party certifiers.

3.5 The MSC Standard

3.5.1 *Standards for certification*

Both the MSC and the group of experts that developed the MSC Standard decided against including aquacultured products in the Standard's Principles and Criteria, for a variety of reasons. Besides additional complexity and important differences with capture fisheries, aquaculture production is also the subject of several efforts to develop guidelines by international agencies such as the UN Food and Agriculture Organization (FAO), industry initiatives, and non-governmental efforts. However, having gained experience in the certification of wild capture fisheries, the MSC received a grant from the Rockerfeller Brothers Fund to evaluate the feasibility of developing an aquaculture certification standard (MSC, 2001b).

The MSC's Principles and Criteria for wild capture fisheries were developed over 18 months. Initial draft principles and criteria for sustainable fishing were developed at a workshop in the United Kingdom in September 1996. The initial draft of the principles and criteria drew upon a number of sources, including the FAO *Code of Conduct for Responsible Fisheries*, the UN *Agreement on Highly Migratory Species and Straddling Stocks*, and statements of principles from several workshops. The draft principles and criteria then were reviewed at eight meetings with a wide range of stakeholders in the United States, Europe, South Africa, New Zealand, Australia, and Canada. A final draft of the *Principles and Criteria for Sustainable Fishing* was prepared at a second workshop of experts in Washington, DC, USA in December 1997, and then forwarded to the MSC Board of Directors. Before final adoption of the principles and criteria by the MSC board, several test cases were conducted, including the Alaska salmon fishery, the Western Australia rock lobster fishery, and the Thames herring fishery.

According to the MSC, these principles and criteria for sustainable fishing are to be used as a standard in a third-party, independent, and voluntary certification programme (MSC, 2001a). Underlying the principles and criteria are several goals:

- the maintenance and re-establishment of healthy populations of targeted species;
- the maintenance of the integrity of aquatic ecosystems in terms of fishing impacts;
- the development and maintenance of effective fishery management systems;
- compliance with relevant local and national laws and standards and with international understandings and agreements.

In the preamble to the principles and criteria, a sustainable fishery is defined as one that:

- can be continued indefinitely at a reasonable level;
- maintains, and seeks to maximise, ecological health and abundance;
- maintains the diversity, structure and function of the ecosystem on which it depends, as well as the quality of its habitat, minimising adverse effects;
- managed and operated in conformity with local, national, and international laws and regulations;
- maintains present and future economic and social options and benefits;
- is conducted in a socially and economically fair and responsible manner.

The three principles are as follows:

(1) A fishery must be conducted in a manner that does not lead to overfishing or depletion of exploited populations and, for those populations that are depleted, the fishery must be conducted in a manner that demonstrably leads to its recovery.
(2) Fishing operations should allow for the maintenance of the structure, productivity, function and diversity of the ecosystem (including habitat and associated dependent and ecologically related species) on which the fishery depends.

(3) The fishery is subject to an effective management system that respects local, national and international laws and standards and incorporates institutional and operational frameworks that require use of the resource to be responsible and sustainable.

Each principle is accompanied by several criteria that provide general guidance on complying with the principle. Principle 1 has three criteria, as does Principle 2. Principle 3 is structured somewhat differently in that there are 17 criteria grouped into two categories. In comparison to the criteria in Principles 1 and 2 (Chapter 4A and B), some of those in Principle 3 are very specific (Chapter 4C).

The MSC does not allow variations in the application of its principles and criteria to specific fisheries (Glowka, 2001). The MSC promotes consistent application through its certification methodology and certification guidelines. The MSC principles and criteria concern fishery activities 'up to but not beyond the point at which the fish are landed'. Issues regarding the allocation of fishery resources and general questions regarding resource access are beyond the scope of the principles and criteria.

3.5.2 Certification

The client for the certification may be a fishing organisation, a government management authority, a processor's organisation, or any other stakeholder. In the case of the Alaska salmon fisheries, the client was the Alaska Department of Fish and Game. The client in the Burry Inlet cockle fishery was the South Wales Sea Fisheries Committee, a regional management body. The Hoki Fishery Management Company of New Zealand, an industry body representing quota holders, applied for certification for the hoki fishery. The client pays for the assessment by a certifying organisation. Costs have ranged between US$20 000 and $100 000.

Fisheries are eligible for certification regardless of their scale. From the beginning, developing country governments and environment NGO as well as small-scale fishermen in a number of countries raised concerns that the cost of certification and requirements for sophisticated management systems would preclude small-scale fisheries and fisheries in developing countries from certification. Sensitive to these concerns, WWF provided technical and financial assistance to several fisheries in developing countries and sought small-scale fisheries in the United States with a view to assisting their application for certification. Several relatively small fisheries in Mexico have recently applied for certification.

Only certifiers accredited by the MSC may assess and issue a certificate for fisheries (MSC, 2001a). In order to be accredited, a certifier must submit an application and pay a nominal fee to the MSC. The MSC's Accreditation Officer then conducts a desk-top review of the certifier's procedures, assesses these against the MSC's accreditation methodology, and issues a report. A critical step is that a certifier has to conduct a fishery assessment and be observed by MSC staff to ensure competence. The MSC Approvals Committee reviews the report and confirms or rejects the accreditation decision. The status of accredited certifiers is reviewed every five years.

The MSC certification process, which has evolved since the first certification in April 2000, begins with a fishery requesting an assessment by a certifier accredited by the MSC (2001a). The certifier assembles an assessment team that must include expertise in fishery stock assessment, ecosystem processes, fishery management, and fishing operations. The certifier and assessment team must contact stakeholders, including fishers, fishing organisations, environmental groups, researchers, and government agencies, amongst others.

The MSC fishery standard is the product of a considerable amount of analysis and discussion amongst some of the world's most senior and experienced fishery scientists, managers, fishers and environmentalists. In addressing both stock issues and issues relating to the impacts of fishing it seeks to implement one aspect of what is still an evolving concept in fisheries – fishery-ecosystem management.

3.6   The expectations

The founders of the MSC sought to create a standard that was applicable to any fishery in the world. The reasons for this are several but included the need to make the benefits of certification available to those that had been successful in managing fisheries as well as to satisfy global trade agreements regarding market access, amongst others.

This comprehensive approach created a number of major challenges, for example:

There is a huge diversity of fisheries in the world. These range in production from small scale, very localised fisheries producing less than one hundred tonnes per annum up to those that produce over one million tonnes of fish per year. There is also a great disparity in the amount of information available to those that manage fisheries and also in the types of information used (e.g. scientific data versus experiential knowledge). Management systems themselves are also diverse and it seems that any type of management system can be successful if the conditions are right. Creating an assessment system which is focused on outcomes rather than preconceived ideas about inputs (e.g. assuming that line fisheries are always better than trawl fisheries or assuming that input controls are better than output controls) is a way of addressing this diversity but it creates the need for evidence that may not be easily available.

Measuring a fishery against the standard inevitably involves making judgements about what is acceptable or not. Like all human activities fishing (whether commercial, artisanal or recreational) has an impact on the environment and there is a huge diversity of strong opinions around as to what is an acceptable consequence of fishing. Not all of these opinions are free from self-interest and many environmental debates in fisheries are rooted as much in allocation claims as any concern about the environment.

Measuring natural systems in order to provide an independent and objective assessment of the impact of a fishery is no easy task. In addition to the many scientific debates about what and how to measure (and what the data mean when they are collected) there is also the challenge created by the fact that a certification assessment is, in essence, a snapshot in time. New information about a fishery is always being generated and there is always a chance that an assessment delayed by a month or a year may be characterised by a different result.

3.7   The result – a broad-based standard

The MSC standard is very broad and not prescriptive in its expectations, although within Principle 3 there are some clear statements about what is not acceptable (e.g. use of poisons and explosives as fishing methods). The principles and criteria establish a series of topics for consideration that have to be further interpreted. The standard for any given fishery is unique for that fishery as each fishery has its own set of characteristics although there are obvious similarities amongst 'families' of fisheries, for example, prawn trawl fisheries or tuna longline fisheries, amongst others. Thus, the creation of a standard against which a fishery is to be assessed is a two-step process – the principles and criteria which are established by the MSC and a series of scoring guideposts and indicators which are established by the certifier's expert team. Both these steps involve public input.

3.8   Overview of how the process works

The MSC certification process is described in detail in Chapters 4 and 5. In order to introduce a description of the fishery assessment system and the issues that arise a short summary is as follows:

Step 1 – a client selects a certifier and a pre-assessment of the fishery is prepared.
Step 2 – if the client wishes, a full assessment is then commissioned.
Step 3 – the certifier prepares draft scoring guideposts and indicators for public exhibition.
Step 4 – the certifier evaluates the fishery against the guideposts and indicators.
Step 5 – the draft report is peer reviewed and a determination is made by the certifier regarding whether or not a certificate should be issued.
Step 6 – prior to a final decision about whether or not a certificate should be issued, the determination may be subject to a formal objection.
Step 7 – once any and all objections have been fully considered a decision about whether or not to issue a certificate is made.

The overall process is characterised by the gathering and analysis of information and the review of the certification team's determinations. The detail of the fishery assessment process reveals the complexity of the task at hand and this is described below.

3.9   Fishery assessment systems

There have been several systems created which evaluate fisheries from a holistic perspective (i.e. that evaluate more than just stock status) and they share some similar basic attributes. Some are based on the FAO *Code of Conduct for Responsible Fisheries* and provide mechanisms for evaluating the performance and progress of a given fishery (or groups of fisheries) against the code.

Caddy (1996) provided a checklist of issues that were determined to describe progress against the code. For most of the attributes (which include stock status, endangered species interactions, management system etc) there is a relatively simple scoring scale (Yes = 1, No = 0). Although the number and scope of attributes could be debated the approach represents a major stepping stone in the creation of such fishery evaluation systems.

Another approach developed by Pitcher (1999) is known as *Rapfish* and is a system for rapidly evaluating groups of fisheries in relation to each other. As an example, Pitcher (1999) compared a group of Canadian fisheries to demonstrate how the system works. *Rapfish* uses a more complex scoring scale than the system created by Caddy.

Whereas *Rapfish* has been subject to scientific peer review it was not subject to broad public consultation and a number of the attributes have very subjective interpretations. For example, attributes relating to income distribution and management type have scoring which is very much culturally based. It should be noted that the MSC standard was originally proposed to have five principles, the fourth and fifth relating to economic and social performance. However, the global consultation process revealed that such aspects were very culturally based and had little clear link to ecological sustainability.

In Australia the federal environment department (Environment Australia) has created a system for evaluating fisheries for a variety of purposes such as the issuing of export permits. This system has a similar structure to the MSC standard in that it has three main principles and a series of criteria but the approach used for evaluating fisheries is based on internal government processes.

3.10   The MSC fishery assessment system

When a fishery enters the full assessment phase the certifier convenes a team of experts, which has the job of creating a series of performance indicators and scoring guideposts for that fishery. The purpose of the indicators is to provide a mechanism for fully evaluating each criterion and the number of indicators within each criterion varies according to what the team feels is necessary to achieve this.

The scoring guideposts provide a mechanism for both converting a subjective judgement into a quantitative score and for making judgements about not only the performance of the fishery but also what constitutes an acceptable performance. Under the current system the performance of a fishery deemed to be perfect determines the upper boundary of the score for that indicator. It is allocated a score of 100. The lower boundary (allocated a score of 60) sets the lower limit of acceptability. A score of 80 is determined to be the boundary between a fishery that is performing acceptably for a particular indicator but one that requires some minor remedial action.

Much has been made of these scores. For example, comments have been received that 60 and 80 are either too low or too high. The fact is that these are arbitrary numbers that define thresholds of performance and it is the narrative description associated with these thresholds that holds the key to evaluating the performance of the fishery.

As mentioned above a draft set of performance indicators and scoring guideposts is released for public comment. The reason for this is to ensure that certifiers do not create a standard that is too low and to ensure that all relevant issues are covered. The team also has the option of weighting the indicators to reflect their importance to the assessment being conducted. Once the final performance indicators and scoring guideposts are determined the certification team is in a position to evaluate the information they have about the fishery. Given the complexity of the assessment process (there may be 50 or more indicators used and some may have more weighting than others) the certifiers make use of proprietary software (*Expert Choice*) which uses the decision support tool called the Analytic Hierarchy Process (AHP) (*www.expertchoice.com*). Each Indicator is given a score out of 100. The AHP calculates a score for each Principle and this score has to be 80 or more for the fishery to pass. With a score of less than 60 at the criterion level the fishery fails to meet the MSC standard.

3.11   Experience to date

In 1999–2000 the MSC conducted three certification test cases and supervised these closely. Two different assessment systems were used – the one now in use was applied to the Western Australia rock lobster fishery and the Alaska salmon fishery, whilst a more traditional checklist approach was used for the Thames herring fishery. The MSC also has experience with three other certified fisheries and partial experience with two of the other fisheries currently in full assessment in that the performance indicators and scoring guideposts have been prepared and, in the case of the South Georgia toothfish fishery, the full assessment has been completed.

The types of issues that have been generated by these experiences have included:

(1)   Redundancy and duplication in the Criteria – sometimes the available information is required to fulfill assessments in more than one indicator.
(2)   Determining the scoring guideposts correctly – some groups have expressed concern that the 'pass' levels have been set too low.
(3)   Software application problems.
(4)   Specific issues related to the format of Principle 3 – the criteria in Principle 3 are too specific in many cases.
(5)   Stakeholder consultation issues – different groups in different countries have varying expectations of the consultation process.

The MSC has responded to these issues in a number of ways. With regards to point 3 a guidance document for certifiers has been prepared by an AHP expert and this will provide clear instructions on the use of *Expert Choice* in the context of MSC fishery assessments. This guidance document is designed to complement a rewrite of the certification methodology (see below) and will be a part of a suite of such guidance documents – one of which will address shortcomings in the stakeholder consultation requirements.

With regards to points 1 and 4 there is an acknowledged need to review the principles and criteria. These are now five years old and it is timely for such a review to be initiated. It is expected that consideration of this task will fall to the newly created Technical Advisory Board which replaced the old Standards Council after the governance review completed in 2001.

Point 2 raises some almost intractable dilemmas. The MSC fishery assessment system is based on expert judgement and is based on scientifically determined facts or other objective and independently verifiable evidence, where appropriate. However, some groups are measuring the performance of the system on whether it delivers outcomes consistent with their campaign or business objectives. Every certification to date has been subject to comments typified by 'We support certification but not for this fishery' and 'If this fishery is/not certified then the MSC is doomed'. The processes of expert judgement and objective assessment cannot deal with this sort of position taking but are fundamental to an independent assessment. The best that such a system can deliver for those that have such expectations is an admission that even if the outcome did not accord with expectations then the process was at least robust, repeatable and fair.

3.12   Current directions

The fishery assessment system is regularly monitored by MSC staff and specific issues are evaluated in several fora, namely:

(1) *Certifiers' workshops* – the MSC runs workshops for certifiers in order to exchange views, canvass changes to the system and educate new certifiers. Prior to the July 2001 workshop MSC staff reviewed the certifications to date and discussed a number of changes to the system and, in turn, acted on requests for greater guidance material.

(2) *The Technical Advisory Board* – the Board provides a source of expertise and advice on strategic and generic issues affecting the fishery assessment process. It has met once to establish terms of reference and procedural matters. Its second meeting in September 2002 will consider some important projects relating to the scope and interpretation of the MSC standard.

(3) *Stakeholder feedback* – the MSC receives comments from a wide variety of people about its programme and specific fisheries and these are evaluated by staff as to their wider applicability to the assessment system as a whole.

Although the MSC ensures that its system is up to date there is a balance between this and not having changes that are so frequent as to confuse stakeholders, certifiers and clients alike. There is also a balance between canvassing ideas with interested parties and conducting lengthy, global consultations on relatively minor issues (see below). Our experience to date has been that the system works basically as planned. However, as with all cutting-edge innovations, there is room for improvement. In addition, there are issues outside the more technical aspects of the existing system that will have implications on its future design and operation. Some of these are described below.

3.13   Some issues of concern for the MSC board, staff and committees

3.13.1   The balance between public interest and timely progress

The vast majority of fish resources are in the public domain and, by and large, the general public has (or should have) a rôle in determining their use and the conditions of such use. In the case of certification, where there are judgements about the acceptability of impacts and potential profits to be made from the use of the MSC logo, it is imperative that input from those not connected with the fisheries sector is allowed for at all levels of operation of the MSC. Mechanisms for obtaining this input are in place at the board level, in terms of MSC policy and strategic direction (via the stakeholder council), at the technical level (via the technical advisory board) and at the fishery specific assessment level. Arguably, the MSC is unique in its range of active stakeholders, its mandate, and the various levels at which input is sought.

All of this has a cost, both financially and in terms of time. It is very easy (and a well tried strategy for some) to tie up organisations in the bureaucracy of consultation and input. Over the past year the MSC has responded swiftly to suggested improvements in the transparency and openness of its operations. Arguably again, the MSC is probably at a stage where any further significant increases in opportunities for input will jeopardise its aim of improving fishery management. This is not only an issue for the board of Trustees but also for the certification teams conducting fishery certifications. Although there is no easy answer, the issue needs to be raised as the MSC needs to avoid gridlock in its fishery programme.

3.13.2   The balance between certifier independence and being the standard owner

The MSC was established to promote independent, third party fishery assessments against a globally applicable standard. The assumption is that the checks and balances built into the system will always deliver fishery assessments that are comprehensive, repeatable and devoid of major flaws. Certifiers take pride in their ability to deliver on this expectation and the MSC has no reason to doubt that this will not continue to be the case. However, if things go wrong, it is the owner of the standard, the MSC, that will bear the criticism. There is already a debate as to how interventionist the MSC should be in the certification process as a whole, the fishery assessment process itself and some specific certifications in particular. There are some strongly held feelings on this type of issue as expressed over the adoption of an objections policy that shifted the timing for permitting an objection to be raised from after a certificate is lodged to before. The MSC proposed minor changes to its process to give it a greater involvement in the certification process without compromising its independence. Where the balance eventually lies has implications for the fishery assessment process as well as the MSC as a whole.

3.13.3   *Definitional issues – aquaculture* vs *wild catch*

The MSC was established to focus on wild-catch fisheries but it has become very obvious that many fisheries straddle the boundary between what seem obvious wild-catch fisheries and what seem equally obvious aquaculture operations. Stock enhancement from hatcheries, stock enhancement from the wild, protecting fish from predators and ocean ranching are just some of the issues where the boundary between wild catch and aquaculture is less than clear. These types of activities are not recent innovations, some of them have been practised for hundreds of years. Since the MSC commissioned a feasibility study on whether an aquaculture standard and certification system should be established, the need for a workable definition of the boundary between the two seafood production systems is urgent.

3.13.4   *Regional concerns – different views about by-catch etc.*

The MSC has encountered significant resistance in some regions of the world for a variety of reasons. Some, such as previous experience with eco-labels imposed from outside, require the passage of time to resolve but others, such as differing attitudes to the use of fish, have strong cultural roots. Although the MSC system has sought to avoid making culturally based interpretations, there is still a perception in some regions of the world (such as Asia) that the MSC has been established to promote particular approaches to fishery management that are based upon experiences in the temperate waters of the Northern Hemisphere. This is particularly exemplified by the debate over trawl fisheries and the use of concepts such as target species and by-catch. In some countries there is no such thing as by-catch because everything is a target. Such fisheries may be challenging to those who view them from a temperate, 'Western' perspective but they are as welcome to subject themselves to certification as any other fishery and, possibly, new evaluation tools will have to be developed to enable assessments to be conducted.

3.13.5   *Costs and the challenges of small-scale and other data-poor fisheries*

A robust and detailed fishery assessment programme, that is paid for on a full commercial basis, can be expensive and this is a significant disincentive for small-scale and developing world fisheries. Although there are some emerging options the cost will continue to be an issue for the life of the MSC.

Small-scale and developing world fisheries tend to have minimal data sets (at least in terms of 'Western' science). There are some serious concerns that the benchmark for what is acceptable is being set by information intensive 'Western' fisheries and that this benchmark is too high. The MSC is working on devising fishery assessment methods for such fisheries that retain the globally applicable standard without reducing expectations of performance.

3.14   Promoting the brand

There is no point in having an eco-labelling programme for sustainable seafood if consumers are unaware of it or unable to buy labelled seafood. This may seem an obvious point but it is sometimes overlooked by some technicians who would happily create the most robust standards, the most scientifically perfect mechanisms and the most democratic processes, leaving the rest of the world to pass through supermarkets and restaurants in blissful ignorance of the issue or its solution.

In promoting a brand the challenges are varied. Firstly, as a charitable organisation, the MSC does not have the marketing and advertising budgets of major corporations at its disposal. Secondly, raising public awareness of MSC products is unwise if it generates an expectation which cannot be met by the current small volume of product. Finally, the MSC brand must condense the complex scientific issues into 'consumer friendly' language.

Although products from all of the six fisheries certified to the MSC standard are on sale at major supermarket chains around the world (with some 150 product lines in 8 countries), take-up of the MSC programme varies enormously. In the UK, the programme is firmly entrenched and all the major supermarket retailers support the programme, with Marks and Spencer, Sainsbury, and Tesco actively working on sourcing MSC products and one in particular promoting the MSC extensively as part of its overall marketing strategy. While in continental Europe the programme has been greeted enthusiastically in Switzerland, this is not yet the case for other European countries. Apart from Unilever's *Iglo* brand New Zealand hoki range (on sale in the Netherlands, Belgium, France and Germany) and a canned salmon line stocked by the Delhaize Bio Square organic store, there are no other MSC labelled products currently on sale in mainland Europe. Take up of the MSC programme in the US is even lower, with only one retailer (Whole Foods Markets) stocking MSC-labelled products. The MSC strategic plan, therefore, seeks to match fisheries to markets as a means of addressing this situation. In other words, where there is demand, the MSC must target species which supply that demand.

The key retail objectives identified in the strategic plan are defined as:

- encouraging leading retailers in target countries to stock MSC-labelled products –
  the Best Environmental Choice in Seafood;

- encouraging their suppliers to source seafood from fisheries certified to the MSC standard;
- encouraging those retailers already stocking MSC-labelled products to take additional species as they become certified;
- encouraging take up of the MSC programme within the food-service sector, initially by leading UK and US restaurant chains;
- maximising logo-licensing revenue for the MSC trading company, MSCI, in order to provide future revenue support for the work of the MSC (see below).

In order to achieve these objectives the MSC programme must be extended from its well-established base in the UK, to other key markets. Mindful of limited resources, and that it is likely to be more successful if these resources are concentrated on areas where there is likely to be most consumer demand, the following regions have been identified as expansion areas: northern Europe, Australasia and North America.

In Europe, Australia and New Zealand the retail sector will be targeted while in the US there will be more focus on the restaurant sector. Research indicates that the countries where there is most likely to be a demand for eco-labelled fish products are Switzerland, Germany, the Netherlands, Belgium and France. This is not to say the MSC will not follow up opportunities that present themselves in other areas such as southern Europe or Japan, rather it acknowledges that with limited resources, success is more likely to be found in targeting efforts on those specific markets.

The US retail sector is extremely diverse (whereas in the UK and other European countries it is highly concentrated) with different chains having market dominance in different areas of the country. In addition to Whole Foods Markets, there is some interest from other natural foods and high-end grocers, and these are pursued as opportunities arise. In the US, over 60% of all seafood is consumed in restaurants and most seafood trends start in so-called 'linen tablecloth' (i.e. up-market) restaurants before moving to retail grocery stores. It was decided, therefore, that initial MSC effort should be targeted on the US restaurant sector. There are also a number of NGO inspired seafood programmes aimed at consumers to encourage 'best choices' in seafood which tie into the MSC programme, e.g. *The Good Fish Guide* (Clarke, 2002).

A number of challenges exist at the time of writing. Firstly, the limited range of certified fisheries currently available. Only two fisheries of real commercial value to large retailers are certified at present – Alaska salmon and New Zealand hoki. Major retailers need to be confident that the range of seafood bearing the MSC label will continue to expand. Furthermore, two of the products in the MSC portfolio, Western Australian rock lobster and Alaska salmon, have relatively high prices compared with competing products, although it should be noted that this higher price is considered to be unrelated to MSC certification. Smaller fisheries such as Thames herring and Devon and Cornwall mackerel are largely of interest to the UK domestic market only. There is an absence at present of high volume certified white fish. Additionally, the seasonality of some of the fisheries currently certified means they are only available for limited periods.

Secondly, there is a risk of adverse reactions to certifications from the NGO community which will inevitably affect retailer buy-in to the MSC programme. For example, the negative reaction to the New Zealand hoki certification by some conservation groups (particularly in Switzerland, Germany and the US), which could potentially influence consumer perceptions, was a source of concern to some retailers and restaurants fearful of an adverse media reaction.

Thirdly, levels of consumer recognition of the problems of fisheries sustainability remain relatively low, although variable depending on which country. Some retailers in the US and continental Europe feel that the issue of fisheries sustainability is not one that concerns their customers. In the UK, however, retailers are sufficiently motivated by the issue to promote the MSC more actively. The outlook in the Australian market is also promising.

The following key species groups have been identified for retail-restaurant sectors in target markets: cold water prawns, warm water prawns, groundfish (e.g. cods and hakes), tunas, crabs, herrings[2] and mackerels[3].

Low volume of product is a risk. That said, a high volume of products which attracted controversy would be equally disadvantageous. It is a chicken and egg problem. The MSC will not undertake a major consumer launch (which would help to address the problems of low-level consumer recognition of fishery sustainability problems) until there is a greater range of products for consumers to purchase. The MSC has specified a target of 10 certified fisheries (at least half of which should be large-volume commercial fisheries) before allocating resources to a consumer launch, to avoid creating consumer demand for a limited supply of products.

3.15   Overcoming the challenges

In the UK, the MSC plans to address this possible lack of high-volume fisheries in two ways. Firstly, encouraging retailers to extend product ranges based on certified fisheries through, for example the development of convenience dishes. There are many different methods of processing and serving hoki or salmon, and thus label proliferation could increase in this way. Secondly, a more detailed assessment of which UK fisheries might be suitable candidates, followed up by direct targeting in conjunction with keen buyers from across the retail sector.

Furthermore, the MSC is actively seeking to improve its relationships with conservation groups that should better enable the organisation to manage reactions to controversial certifications. (These issues are covered elsewhere in this chapter.) Most importantly, however, is the rôle existing retail supporters can play in applying pressure on their suppliers to persuade fisheries from which they source to come forward for certification. A recently established European commercial group will be an important forum in which retailers and processors can work together to send a

---

[2] Although we do have a source of certified product from the Thames herring fishery (UK), the season is very short (November–March) and volumes are small.

[3] We also have a certified mackerel fishery from the Devon and Cornwall (UK) handline fishery but again, volumes are low.

clear message out to fisheries that they are actively seeking to source from fisheries certified to the MSC standard. Additionally, more direct communications selling the market benefits of MSC certification to the fisheries sector are required.

3.16   Paying for the change

Attracting funds to take on the global task of providing real incentives in the marketplace for better fishery management has been a long and difficult process. Initially, when the MSC was formed in the mid 1990s, some assumptions were flawed. It was envisaged that logo licensing revenue to the charity would quickly fund the costs of the organisation, a bold prediction indeed. Some even thought that by 2000 the MSC would be entirely funded through logo licensing. How wrong they were.

In 2000, the MSC still struggled to attract more than around £350 000 to fund its mammoth programme. Perhaps more significantly, that money came from only three or four sources. Accusations of being in one pocket or another were inevitable. The two founders, Unilever and WWF, were still perceived to be bankrolling an operation that was largely there to serve their interests at the expense of others. That was never the case, but perception is always more important than reality. The MSC, like so many embryonic charities, faced constant danger of financial collapse and loss of funder confidence. By 2002, however, the picture was radically different. The four early funders had become a large family of over fifty. A wide range of charitable trusts and foundations, corporations (including retailers, processors, banks, restaurants, airlines, cruise lines, and a range of 'in kind' or *pro bono* donors) and international development agencies now assist the MSC. Some provide much-needed, core running-cost funds, others prefer specific sponsorship of events, human resources, marketing materials, programmes, or one-off requirements. The MSC is also attracting partners for cause-related marketing initiatives. By 2002 annual income was somewhere in the region of £2 million.

This much rosier picture is, however, deceptive. Firstly, the MSC cannot do what it does without attracting £2 million per year. Secondly, to a large extent the improved fundraising initiatives took place against a backdrop of relative global economic stability but was showing signs of weakening. No one has tried raising funds for sustainable fishing during a world recession. Thirdly, is the age-old problem of what is sometimes termed 'donor fatigue', donors seldom stick to one issue indefinitely. Many like to provide start-up funds and then move on to new pressing issues. Indeed, the MSC was to some extent a beneficiary of a move from forestry to fishery issues in the mid 1990s.

By the end of 2001 it was becoming clear, following the global political and economic turmoil in the run up to and after September 11th, that priorities and focus were shifting for some corporations and philanthropic organisations. This did not pose an immediate danger for MSC, but these are uncharted waters. The importance of the MSC trading company (MSCI, a subsidiary commercial entity 100% owned by the charity) as a source of income is considerable. In 2001 there was a significant level of activity for MSCI as it embarked upon the process of developing the logo licensing system and its business strategy in order to fulfil its primary mission – to generate sufficient revenue to make a significant contribution to MSC core costs. The initial objective is for MSCI to cover its own costs by 31st March 2004 and to be making a net contribution to MSC thereafter. The number and volume of MSC certified fisheries are critical factors. It is crucial, therefore, for the MSC to engage high-volume, commercially valuable fisheries in its programme without compromising its environmental standards as a means to that end.

The work of MSCI is largely divided up between logo licensing, business planning and business development. There has been a steady increase in the number of organisations applying to use the MSC logo on- and off-product as products from the 6 certified fisheries flow through the market place and information about the MSC becomes more widespread. By December 2001, approximately 27 on-product agreements had been signed by 19 different organisations wishing to display the MSC logo on packaging containing seafood from fisheries certified to the MSC standard. It is anticipated the number of licences issued will continue to increase as will the revenue they generate. Approximately 145 off-product licences have been issued for the display of the MSC logo. As this logo becomes more sought after and more valuable, the cost of using it, and hence the revenue to the MSC, will increase.

Accreditation fees are also a potential source of greater revenue. MSCI currently charges certifiers much less than other accreditation bodies. This area is therefore under consideration. There may also be possibilities in the area of grants, subsidies and sponsorships, perhaps from governmental sources. Additionally, the fields of training, general consultancy, and goods and merchandising are all areas for consideration. Funding the MSC remains a major challenge and one that remains at the top of priority lists in order that the organisation can continue to meet the expectations of its stakeholders.

3.17   Conclusions

The MSC, by its very nature, is an evolving organisation. It is seeking to tackle a pressing environmental problem for which there have been few lasting solutions. Independent certification of fisheries has never been attempted before on a global scale and the task of building an ever-growing base of consensus on how to achieve real incentives for improved management is never ending. The very concept of eco-labelling itself continues to be the subject of debate, whether in the World Trade Organization, European Union, FAO or NGO sectors, amongst others.

3.17.1   *Fishery science is not perfect and nor is the MSC programme*

For all its obvious merits, the Marine Stewardship Council is not uncontroversial. How could it be when it is seeking to influence one of the most controversial fields in environmental stewardship? The global fishing industry and its sometimes tense relationship with conservationists, retailers and governments is, by definition, a highly politicised and fractious sector. Some conservationists are as sceptical of the MSC efforts as even the most hard-line industry representative. Again, what is at stake is the proposition that there might be a new and better way of doing business. The MSC, as an accreditation and standard-setting body, must sit precisely in the middle between all the conflicting ideologies, priorities and personalities with whom the fishery world is peppered.

A useful analogy may be found in the field of politics when a dramatic shift in public opinion leads to the election of a new party or leader on a platform of sweeping change, a new style and a challenge to entrenched beliefs across society left by the previous government. French Gaullism, American Reaganomics, UK Thatcherism or more recently British New Labour could all be said to fall into this category. Such movements are magnets for disillusionment and by nature attract visionary, futuristic thinkers of the time. Having done so, the challenge begins when the government realises that keeping an unholy alliance of stakeholder support together is almost impossible at times. This is because the stakeholder alliance is not a natural coalition, rooted in history. The age-old cleavages have been broken apart. Voters who joined the new movement out of frustration with their 'natural' party have to be kept happy if the government is to survive. However, often they are kept happy at the cost of some traditional supporters who have always supported the political party or leader, who feel their own representatives have sold out, watered down, or compromised principles too far. What results is often a rather vague, bland and ultimately unsuccessful regime, as the new alliance finds itself unable to respond adequately to economic crises, domestic political scandals or other shocks to the system. Voter loyalty is shattered far more quickly than it is gained. Any government that is born out of a desire for radical change fuelled by discontent faces this precipice. One concession to the left wing, and the right feels betrayed and *vice versa*.

What relevance is all this to the MSC? In many ways, the MSC position is remarkably similar. To some environmentalists, the MSC is the more or less acceptable face of industry, or at least the industry that has chosen to work with the programme. To many in industry, the programme is the acceptable face of environmentalism. Our stakeholder tent must accommodate fishermen, fishery managers, governments, retailers, processors, restaurateurs, conservationists, the scientific community, certifiers and ultimately, the public. The MSC eco-label must be sought by consumers, valued by industry, endorsed by conservationists, respected by governments and practical for certifiers. In a field where there remains little global consensus on how to bring about a real sea change in fishery management, one tiny concession to one group at the expense of another can have damaging political consequences for the programme and its credibility. Early lessons suggest that if the MSC acts rapidly on the demands or whim of one group, it is often to the detriment of relations with others. On the other hand, it cannot be seen to be inflexible, unaccommodating and static. This dilemma is not dissimilar to those of the political scene described above. Keeping this bubble together, united in purpose and afloat is a constant challenge, and success has been mixed.

3.17.2 The rôle of good governance

One method of alleviating some of this pressure lies in a robust governance structure which is open and transparent, but not to the extent that decision-making processes are paralysed. The MSC reviewed its entire system in 2001 and commissioned an independent panel to revise the existing structure with these aims in mind. The previous structure had become excessively bureaucratic and complex, with so many rules, committees and requirements that some of the bodies attached to the MSC had never met in person or even been appointed. Those bodies that did exist were appointed entirely by the MSC itself, hardly a recipe for openness and transparency.

The answer, however, did not lie in creating a vast democratic organisation in which every decision was put to a vote, which tied the MSC board to decisions with which they did not agree and made progress impossible. That would have been a step too far. In the end, it was decided that two main bodies should exist. The first, a stakeholder council, half appointed by MSC, the other half appointed by the group with no MSC involvement. This group would be able to advise, express concern, deliberate on current challenges and indeed feed directly into the MSC's most senior policymaking body, i.e. the MSC board. The second, a technical advisory board, with a similar mandate although focussed much more specifically on the technical and scientific aspects of the MSC standard and its application. Crucially, the previously existing gulf between the MSC board and all these stakeholders will now be bridged in a very simple but effective way. The stakeholder council (which is evenly divided into 8 constituencies of 5 members from every sector with whom the MSC interacts) has appointed two co-chairs, one from those group of constituencies which fall into the public interest sector (e.g. NGO and other non profit-making bodies), the other from the commercial (i.e. profit-making) sector. Both these co-chairs have been co-opted to the main MSC board. In this way, the stakeholder council will have strong representation at board level. Similarly, the MSC board has also co-opted the chair elected by the technical advisory board, thus giving better scientific and technical knowledge representation at board level. In this way, the board does not become completely dominated by these groups, but by having three trustees who are, in effect, 'the people's choice', the MSC structure becomes more open and its board less remote. Furthermore, because the MSC no longer appoints all these committee members, there is more room for criticism to be absorbed into the system. Time will tell, but the structure appears to have attracted endorsements from across the spectrum, not least because a large-scale consultation exercise was conducted prior to its creation. In this field, the MSC kept its coalition together. The process demonstrates the benefits of good stakeholder consultation before making binding decisions.

At the time of writing, the new bodies have held their inaugural meetings. It is perhaps too early to draw lasting conclusions, but the principle of greater openness and transparency has been established. Few would argue that the MSC could survive without it.

Introduction to the MSC Certification A: The Process of Certification

4

Chet Chaffee

4.1 Introduction

Under the MSC programme, a labelled seafood product with an MSC logo means that both the fishery management and the processing operations associated with the fish in the product have undergone an independent evaluation and certification by an accredited certification body (company). Any company interested in becoming an accredited certification body under the MSC can apply directly to the MSC. The process of certification is described here and is followed by an account of the three key principles that underpin the certification process.

4.2 Fishery certification

When a fishery decides to participate in the MSC programme, it first selects an accredited certification body to provide the independent assessment. The MSC provides a list of accredited certifiers on its website (*www.msc.org*). A fishery evaluation is a two-phase project that first involves a pre-assessment of the fishery and then a full assessment.

A client, who can be anything from a private company to an industry association or a government agency, interested in moving forward toward certification first contracts with an accredited certifier to conduct a pre-assessment of the fishery or fisheries of interest. The pre-assessment under the MSC programme is similar to a feasibility study or a 'gap' analysis found in other programmes such as ISO. Using readily available information, a snapshot of the fishery is compiled and an analysis performed to determine the strengths and weaknesses of the fishery compared to the MSC standards. In addition, the document contains an estimate of the costs associated with conducting a full assessment. The document is confidentially provided to the client and not publicly reported. The client uses this document as the basis for making a decision about continuing to seek certification for the fishery.

Once the decision has been made to go through a full evaluation, the certifier will begin the process of assessing the fishery fully against the MSC principles and criteria. The process is open to the public and involves consultation with interested stakeholders, which includes any interested party from industry to conservation groups, government agencies, or individuals. The end of the process includes the publication of a report documenting the findings of the evaluation team and the recommendation by the certifier as to whether the fisher or fisheries should receive certification (see Chapter 5 for a detailed description of the MSC certification process).

4.3   Chain-of-custody certification

Once a fishery is certified, a chain-of-custody for the products manufactured from certified fish must be evaluated and certified. A chain-of-custody evaluation and certification provides proof that any product sold under the MSC logo or eco-label can be shown to originate from a certified fishery. The chain-of-custody evaluation and certification must be applied to all entities involved in the supply chain from the fishery to labelled product. This includes primary processors, secondary processors, wholesalers, distributors, importers, retailers, food service, restaurants, or any other business that handles MSC product.

B: Principle 1 – Stocks

Louis Botsford

4.4 The guiding principles and criteria

4.4.1 Principle 1

A fishery must be conducted in a manner that does not lead to over-fishing or depletion of the exploited populations and, for those populations that are depleted, the fishery must be conducted in a manner that demonstrably leads to their recovery.

4.4.2 Intent

The intent of Principle 1 is to ensure that the productive capacities of resources are maintained at high levels and are not sacrificed in favour of short-term interests. Thus, exploited populations would be maintained at high levels of abundance designed to retain their productivity, provide margins of safety for error and uncertainty, and restore and retain their capacities for yields over the long term.

4.4.3 Criteria

Principle 1 focuses on the population status of the targeted species, specifically on how to assure that the fishery is conducted so that the population can continue to persist. The principle itself is simple, the fishery should not lead to over-fishing nor depletion, but if it does, the fishery should allow recovery.

(1) The fishery shall be conducted at catch levels that continually maintain the high productivity of the target population(s) and associated ecological community relative to its potential productivity.
(2) Where the exploited populations are depleted, the fishery will be executed such that recovery and rebuilding is allowed to occur to a specified level consistent with the precautionary approach and the ability of the populations to produce long-term potential yields within a specified time frame.

(3)   Fishing is conducted in a manner that does not alter the age or genetic structure or sex composition to a degree that impairs reproductive capacity.

The statement of intent fleshes out the simple Principle 1 goal somewhat. It identifies an important aspect of the problems faced: pressure for high catches and profits, at the expense of long-term returns. It also identifies uncertainty as a key element of the problem, and introduces the idea of maintaining a margin of safety because of that uncertainty. Pressure for higher catches and uncertainty in predicting their potentially deleterious effects in the long run frequently lead to constantly increasing fishing effort, a mechanism referred to as the ratchet effect (Ludwig *et al.*, 1993, Botsford, *et al.*, 1997, Botsford & Parma, 2002).

The first two criteria more or less repeat the principle, calling for (1) the maintenance of high productivity, and (2) rebuilding when depleted. The second criterion also introduces the precautionary approach, at least for recovered populations. The third criterion addresses the fishery impact on population structure in terms of age, genetics and sex. It requires that they not be changed to the degree that the fishery degrades productivity.

These are laudable goals, and few would disagree with them as goals, however, they are of limited usefulness operationally. In fact, there have been few, if any, fisheries that did not include long-term productivity implicitly as a management goal, even those that have collapsed. The well-known problem is that having this goal is not sufficient to prevent collapse. The efficacy of adopting maximum sustained yield (MSY) as a goal for fisheries was challenged as long ago as the late 1970s (Larkin, 1977). In this problem with such a high degree of uncertainty, and other complicating factors, simply striving for MSY often led to population collapse. That one should allow a population to recover following collapse is also an approach on which there would be wide agreement. However, the problem in fishery management has been how to make the management decisions that the fishery is over-fished, and how drastic the recovery actions should be.

That a fishery should have a minimal impact on age, sex and genetic structure is also inherently desirable, but again requires further elaboration. Taking age structure first, it is important to realise that a population cannot be fished without changing the age structure. The question is, when is that change large enough to be dangerous? With regard to sustainability, or population persistence, changes in the age structure affect the equilibrium level of population abundance and the ability of the population to withstand variability in the environment. A critical population parameter is lifetime egg production, LEP (Botsford & Parma, 2002). LEP must remain greater than a minimum value for the population to persist. This is similar to the notion that we can keep human populations from increasing by requiring that each couple have two children or less (i.e., as in zero population growth, ZPG). The difference is that we can calculate how LEP is reduced when fishing culls individuals before they reach older ages. However, we do not yet have a firm idea of what the minimum LEP required for sustainability is for fish populations.

From a population point of view, reducing genetic variability within a population increases the risk that the population's natural ability to persist is compromised.

From a more general conservation point of view, it also reduces biodiversity. The sex ratio is sometimes distorted in finfish fisheries, particularly those with size limits and sexual differences in size and growth rates. However, one must also keep in mind that in a number of fisheries, in particular crustacean fisheries, management allows only males to be fished, in a deliberate attempt to maintain high reproductive capacity. While this is inherently safer from the point of view of sustainability, it can lead to maintenance of high densities and density-dependent effects that may not be desirable for productive fisheries.

4.5   Approach to assessment

Assessment of whether a fishery is likely to achieve the MSC goals requires going beyond the principle, intent and criteria. This has been done in the development of scoring criteria and guideposts by certification teams addressing individual fisheries. In preparing this, I have examined the scoring criteria and guideposts for Australian rock lobster, hoki, Alaska salmon and British Columbia commercial salmon fisheries.

The most important factor to include in developing scoring criteria and guideposts for a specific fishery is a means to assess whether management of the fishery accounts for uncertainty, seeks to quantify it, and is generally robust to its effects. The approach should be more sophisticated than simply allowing a safety factor; rather it should actively seek to minimise the potential deleterious effects of not knowing critical aspects of the population dynamics of a species.

Most of the available instruments for reducing the sensitivity of fishery management to uncertainty can be found in the FAO precautionary approach to fishery management (FAO, 1996). Probably the most important tactic in the context of sustainability is the addition of a limit reference point, in addition to the usual target reference point (Caddy & Mahon, 1995). The target reference point (TRP) is a goal such as the MSY, while a limit reference point (LRP) is not a goal, but rather a *pre-agreed* point at which the population will be declared to be overfished, and draconian measures will then be taken to recover the population. There are often measures in place actively to avoid exceeding the LRP as it is approached. Experience has shown that without a LRP, management bodies will continue to seek their target, even though the populations remain at low levels, or continue to decline. The existence of a LRP, separate from a TRP is essential for sustainability. Fortunately, LRPs are a feature of the management system in most of the fisheries certified at the time of writing.

There are many forms of limit reference points depending primarily on the data available. Ideally, they should be based on a logical connection to population persistence and available data. Examples of LRP include a certain fraction of unfished biomass (e.g. 20%), a certain fraction of lifetime egg production, and the fishing mortality rate that would produce MSY. Population persistence depends on population growth rate and variability therein, as well as current population abundance. As such, LRP should make use of available data to calculate these.

A second requirement is a comprehensive statement of how management is to be conducted. Ideally this is a statement of what action will be taken for every possible observed state of the fishery. A concern with regard to sustainable fisheries is that this statement contains an LRP in some form. This is sometimes drawn out graphically as a decision rule. A pre-agreed, clearly stated approach to management, based on the best available science, is essential to sustainability because it leaves less room for avoiding difficult but necessary steps that may require sacrifice of short-term benefits (e.g. reduction of short-term fishing capacity). The existence of such a document is also necessary to assess and certify sustainability.

Development of the management strategy or decision rule depends on there being a robust assessment of the stock, another common requirement for certification. Achieving robustness to uncertainty commonly involves identification of dominant uncertainties and analysis of the sensitivity of decision rules to them. The decision rules should be structured such that the occurrence of an error in parameter estimation should not allow the population to become overfished.

Identification of the dominant sources of uncertainty should include assessment of a particularly insidious one, environmental variability on slow (decadal) time scales. Estimation schemes used in stock assessment in fisheries can be designed to cope with year-to-year random variability in the environment, but it is very difficult to detect and remove the effects of slow changes in life-history parameters such as growth, reproductive and mortality rates. The consequent errors can lead to errors in the choice of critical elements of management such as LRP. There is an increasing appreciation for the presence of decadal scale change in the ocean and certification should require that managers are aware of their potential effects in a fishery. The salmon fishery in Alaska is a good example: catches increased dramatically following well documented physical and biological changes in the Gulf of Alaska in the mid 1970s. This suggests that there should be a plan in place for detecting and accommodating a reversal in ocean conditions.

Another element that commonly appears in the scoring criteria is the existence of sufficient information on the population and the fishery to manage the fishery adequately. This commonly requires that spatial structure, abundance, productivity, and vital rates, such as growth, mortality and reproduction are known. It is important to know the geographic extent of the population being managed, so that the effects of removals on population dynamics can be defined. However, while knowing life-history characteristics certainly reduces the uncertainty in management, and increases the options in tactics, knowing all of them is not an absolute requirement for sustainability *per se*. Knowing them is not the important point; rather, what you do with them is important (see Peterman, 2002).

4.6   Potential problems, future directions

While certification has got off to a reasonably good start, as more fisheries are considered, specifically different kinds of fisheries, some potential problems can be anticipated. Low data availability will present a problem, both a technical problem, and a problem of fairness. Most of the considerations regarding certification presume a substantial database and fishery infrastructure. That database may not exist even in developed countries. This leads to the unanswered technical question of how sustainably fisheries can be managed with few data. More specifically, are there management measures that are inherently precautionary and robust to poor data or lack of knowledge? There is also an economic issue regarding whether it is worthwhile gathering the data necessary to assure a sustainable fishery, and what to do instead. Developing countries rarely have the fishery infrastructure that other nations have, and some means will have to be found to give them equal access to certification without reducing the fundamental requirements.

Multi-species and multi-population fisheries will also present problems for certification. The considerations above of how to abide with Principle 1, indeed Principle 1 itself, is focussed on a single population. If the fishery is by nature multi-species, or consists of many populations of the same species, that will raise problems. Certification of the salmon fisheries in Alaska raised the multi-population problem. The salmon populations in Alaska consist of 5 species and approximately 30 000 individual spawning populations. The approach taken there was to note the well known problems associated with managing multiple populations with aggregate data (e.g. Hilborn & Walters, 1992), then to require that as many populations as economically possible be directly monitored and managed to maintain sustainability.

A problem that will arise in all fisheries is the potential impact of where the burden of proof of sustainability rests. Keeping costs of certification at an affordable level, so that it will be widely used, requires that the burden of demonstrating sustainability rest on the body requesting certification. There can be a tendency for that body to claim that their fishery is sustainable, and then challenge the certification team to show that it is not. It would be prohibitively expensive to operate certification in this fashion, however, since that would require the certification team to do the management analysis necessary to assure sustainability. The burden of demonstrating that the fishery is sustainable must rest on the management body for the costs of certification to be affordable. Another cautionary note is the observation that it is to the advantage of the requesting body to negotiate as low a price as possible, thus reducing the amount of analysis that can be done by the certification team. As costs are driven too low, the quality of certification will suffer.

Lastly, it appears that at this point the certification process could be streamlined and made more uniform by adopting general scoring criteria for Principle 1. As each fishery comes up for certification, each new certification team is, to a degree, reinventing the wheel. The cost of that redundancy could be avoided if general criteria were adopted. Such a step would also assure greater uniformity among the different certifications. The only potential problems with doing this would be that they might limit the innovations that will be required as data-poor and multi-species situations are considered.

C: Principle 2 – Effects of Fishing on the Ecosystem

Trevor Ward

4.7 The guiding principles and criteria

4.7.1 Principle 2

Fishing operations should allow for the maintenance of the structure, productivity, function and diversity of the ecosystem (including habitat and associated dependent and ecologically related species) on which the fishery depends.

4.7.2 Intent

The intent of this principle is to encourage the management of fisheries from an ecosystem perspective under a system designed to assess and restrain the impacts of the fishery on the ecosystem.

4.7.3 Criteria

The criteria for Principle 2 cover the key aspects of ecosystems and the range of possible impacts of fishing. An MSC assessment would normally consider all aspects of a fishery's possible impacts on the ecosystem, and all elements of associated eco-systems that may have interactions with the fishery.

(1) The fishery is conducted in a way that maintains natural functional relationships among species and should not lead to trophic cascades or ecosystem state changes.
(2) The fishery is conducted in a manner that does not threaten biological diversity at the genetic, species or population levels and avoids or minimises mortality of, or injuries to endangered, threatened or protected species.
(3) Where exploited populations are depleted, the fishery will be executed such that recovery and rebuilding is allowed to occur to a specified level within specified

time frames, consistent with the precautionary approach and considering the ability of the population to produce long-term potential yields.

The intention of the assessment is to determine the extent to which the fishery has identified, addressed and resolved all the relevant and important impacts, or if they have not been resolved, the extent to which the fishery has appropriate processes for identifying the potential issues, and proceeding to resolve them. This will include monitoring procedures to ensure that the fishery keeps track of environmental conditions, and learns of any environmental matters that may arise unexpectedly.

Criterion 1 is focussed on assessing the extent to which a fishery disrupts the normal functions of the ecosystems where it operates. Specific matters of interest are changes to the normal trophic relationships of species in the ecosystems, and particularly if these create additional stresses on species over and above those that would normally apply in a non-fished ecosystem. A crucial aspect of assessing this criterion is the extent to which normal functions and species relationships in the ecosystem can be determined. This is probably best assessed from comparative studies conducted in non-fished ecosystems. The standard expected of a well-managed fishery is that there would be only minor changes, if any, to species and trophic relationships and ecosystem status. Determining the nature and extent of such changes is likely to be very difficult for most marine ecosystems and fisheries, and the MSC assessment process would normally accept as satisfactory evidence of good management that there is an active process of identifying such ecological issues and activities to resolve any such issues are effective and ongoing, including relevant focussed monitoring, scientific research projects, and procedures to correct any specific impacts determined to be of ecological importance.

Criterion 2 is focussed on assessing the extent to which the structure of ecosystems is affected by the fishery. Specific matters of interest would include the extent to which the fishery affects the species richness and composition of ecosystems, and the distribution and abundance of individual species of non-target organisms. Also of interest will be the potential for the fishery to interrupt normal gene flow and evolutionary responses to typical environmental pressures, and to distort the natural (unfished) age- or size-class structure of populations. The effects of the fishery on the natural mortality of important species, or the interruption of their normal behaviour is also a matter of interest for an MSC assessment. These matters are considered most crucial in the case of any species that are considered by any competent jurisdiction (such as the national government, or the International Union for the Conservation of Nature – IUCN) to be a species that is formally declared to be threatened or endangered, or is otherwise formally protected for any specific purpose, such as in a population recovery plan. The ecological importance of all of these are assessed in terms of nature, extent and magnitude of the impact of the fishery – the MSC process assesses how substantive these ecological effects might be in terms of the genetic, species and population characteristics of ecosystems.

Criterion 3 is focussed on the extent to which the fishery must rebuild any depleted populations of the target species. The procedures and approach of the fishery to rebuilding populations is assessed in terms of conservative harvest strategies and ecological impacts on associated and dependent species. Where stocks are not heavily depleted, and there is no evidence of gross overfishing, this criterion is not used in an MSC assessment, and does not contribute to the scoring procedure under Principle 2.

4.8   Approach to assessment

The assessment of environmental performance of the fishery is concerned primarily with the fishery as a whole, because this is the target for any MSC certification, but will consider the contributions made by all stakeholders to the fishery management system. Thus, where the fishery management is conducted by a combination of government and private sector organisations, it is the combined product of all those involved with the fishery (including stakeholders) that is assessed when considering the environmental performance of a fishery. This means, for example, that where a government agency monitors and manages seal populations and a fishery has an incidental catch of the same species of seal, a seal monitoring programme may be considered to be contributing to the management of the fishery by providing data and information of value to the fishery, even though such a programme may be designed and funded entirely separately from the fishery. Of course, if such a monitoring programme did not adequately monitor seals in a way that was useful for the fishery, for example if the monitoring data did not adequately resolve changes in the seal population, then this monitoring would probably not be considered to be adequate for the MSC assessment purposes. Similarly, where a government agency implements some aspects of the fishery management system, such as setting broad policy and strategy parameters such as a total allowable catch (TAC) for the fishery as a whole, and the fishery itself decides on operational matters, such as place, time and gear types for fishing, then both sets of decisions and activities contribute to the management of the fishery, and will fall within the domain of the MSC assessment, since both can profoundly influence environmental performance and sustainability of the fishery.

Where a fishery exploits only part of the normal range of a species, the MSC assessment will consider the condition of the whole population, including those parts outside the fishery being assessed, in order to determine if the fishery is well managed and sustainable. This is to ensure that the fishery being assessed takes full account of other sources of stress (such as harvests by other fisheries, or the effects of pollution on stocks or juvenile habitat) in setting the levels of allowable catch, or other controls, and considers any interactions of these matters (externalities) with the fishery and the part of the population being fished. The absence of reliable information about sub-populations, stocks that are fished in other fisheries, or the impacts of other external factors on the target species that are outside the fishery being assessed cannot be used as evidence of good management. The assessment will normally consider evidence that the externalities are fully considered when determining management strategies in the fishery being assessed, and issues are appropriately considered and reflected in conservative management decisions.

The assessment considers the knowledge base within a fishery to determine if there are any specific matters of environmental concern. The experience of the assessment team together with stakeholder and fishery inputs are used as the basis for determining if the fishery has appropriately identified and addressed any specific issues. Where issues have been identified and resolved, the assessment will normally examine the nature of monitoring and awareness systems to determine if there is an adequate basis for maintaining a current awareness of any known or potential environmental issues. This monitoring system may, and often will, include monitoring and awareness systems operated by other agencies, and may be shared with other fisheries, or operated for ancillary reasons, such as public health programmes.

Where a fishery has not identified any specific environmental issue, but has not considered the possibility of such impacts occurring, then the assessment will consider the nature of any existing processes that may be in place that could lead to the identification of environmental issues, and their potential scope, effectiveness and timeliness. This may include the activities of government agencies other than a fishery management agency, community groups or NGO involved with environmental issues, or the processes in other perhaps related fisheries. Progress towards environmental awareness and responsiveness will be considered in the assessment, and while such activities are not in themselves evidence of sustainability, such responsiveness must be present in order to identify the flexibility and adaptiveness of the fishery to deal with such issues if they arise. Evidence of the effectiveness of such processes is a normal component of the assessment process. This responsiveness is considered an important feature of a well-managed fishery.

In an MSC assessment the performance of the fishery in the context of its environmental impacts (Principle 2) can be the major, and determining, matter of certification. Many fisheries have devoted substantial financial and human resources to developing and implementing high quality stock management systems, because this has a direct bearing on the continued viability of the fishery in the sense of being able to continue to catch fish from an abundant stock. However, few fisheries have devoted a major effort to dealing with environmental issues where such issues have not been linked to direct stock maintenance matters. For example, fisheries habitat reserves are often designed and implemented to protect juvenile stages or spawning aggregations of a target species. While such reserves will make a potentially important contribution to broader biodiversity conservation goals, their purpose is primarily for stock protection and management.

Nonetheless, environmental issues that do not necessarily directly affect the stock are still likely to be important issues for a fishery to examine and resolve. This may be important, for example, where a fishery affects the food supply of highly valued but non-fished species such as seals or other marine mammals, or where a fishery affects the natural feeding patterns of seabirds by providing an artificial food source through discards from fishing catches. Beyond these single-species issues, the assessment must fully consider the non-commercial species of the habitats and ecosystems being fished. Here, the impacts of fishing will be difficult to determine, but such knowledge gaps are not a basis for ignoring the possibility of such impacts.

Despite the potential importance of these Principle 2 issues, many fisheries will not have an adequate amount of the relevant data or knowledge properly to inform an

MSC assessment. This may be because the fishery is inadequately prepared for the assessment and has not assembled the appropriate information for the assessment team, or it may be because the relevant data and knowledge does not exist. For example, much data that is relevant to an assessment of the impacts of a fishery may be held by other agencies, such as conservation or environment protection agencies, and it may not be readily available to an MSC assessment team.

For some matters, such as the ecological effects of a fishery on species that are not themselves fished but co-occur with the target species and are ecologically linked in the form of prey or predator, most fisheries will have only limited data. Also, the amount and type of data that will be needed satisfactorily to determine the nature and extent of fishery effects is likely to be very extensive, and very costly for a fishery to collect. While such matters may well be issues for stakeholders, determining the level of impact of the fishery may take many years of research, and depending on the type of impact, may be best approached in an incremental manner. Here, in a progressive and incremental manner, a fishery would take a carefully planned approach to resolving the issue by conducting original research, but joining in partnerships with other agencies and stakeholders to identify and promote the need for such research, and as appropriate influencing research priorities in other sectors. It is crucial, however, that the assessment does not simply gloss over these matters as 'too hard' because they will always form a core concern for environmental groups who, correctly, consider sustainability to include all the species and habitats where a fishery operates, or might affect.

4.8.1   Assessing effects on biodiversity

The wording of Principle 2 is clear: the assessment focus is on diversity and function of the biological aspects of ecosystems that are linked to the fishery. This includes all aspects of the pelagic and benthic environments, as well as terrestrial areas that may have a direct connection to the fishery, such as for example, where there is a by-catch of mammals or seabirds near important nursery grounds or rookeries, even if such species are not threatened or endangered. However, the MSC assessment is based on determining acceptability of impacts, since it is clear that fishing must have some level of ecological impact, and most of all on the harvested species. In this sense, a well-managed fishery is not deemed to have no ecological impact, rather it will have a level of impact that is consistent with the best managed fisheries of its type, and the effects will be generally limited in scale and be of an extent and nature that is broadly acceptable to most stakeholders.

4.8.2   Performance indicators

Interpreting the broad guidance of the words of Principle 2 into the operational interpretation that can be assessed and scored within the assessment process is a crucial aspect of the MSC assessment procedures. Developing the operational interpretation will always involve adopting some measure of surrogacy in order to represent the broad biodiversity concepts within a small set of measurable parameters that are relevant to the fishery. The performance indicators are the operational expression of the MSC principles and criteria, and are normally customised for each fishery being assessed in order to ensure that the scoring process is not overly biased by the use of generic performance parameters that do not apply to the fishery in question. The effect of such bias may be either to increase or reduce the scores of a fishery, but in either case, a poorly chosen set of performance indicators would not permit an accurate assessment of the fishery, opening the outcomes to inaccuracy and stakeholder dispute. A preponderance of weakly framed performance indicators can outweigh the scoring on one strongly worded indicator and lead to a fishery perhaps falsely becoming certified simply based on the construction of the indicators rather than on specific aspects of its performance.

Choosing a set of performance indicators that are appropriate for a fishery but that are neutral in terms of scoring bias is a highly specialised task, and great attention must be given by the assessment team to ensure that the performance indicators cover all the relevant environmental issues without undue focus on issues where the fishery might perform either well or poorly. A set of indicators that do not apply to a fishery will create an ineffective assessment, even though they may apply to the specific MSC criteria, and potentially will overly reward a fishery by developing a focus on irrelevant matters. Equally, a set of indicators that focus wholly on matters thought to be of contention in a fishery will create an ineffective assessment, and will most likely overly penalise a fishery and fail to give credit for previous environmental achievements. The need for an unbiased and properly balanced set of indicators is paramount if the assessment is to be objective and defendable. It is here where inexperienced assessment teams, fisheries and stakeholders can find the going tough, and in the first instance they may not even have enough experience to be able to recognise the difficulties inherent in the various facets of 'criterion failure'.

The performance scale for each indicator is established by the assessment team with reference to the MSC principles and criteria, and with regard to established environmental practice in other fisheries and other sectors. Normally, the indicators will relate to the specific terms of each criterion, appropriately customised so that they have a direct relevance for the fishery being assessed. The scoring guidelines for each indicator are developed to provide guidance for the assessment team in determining the level of performance and score of the fishery on each indicator.

4.8.3   *The burden of proof*

An MSC assessment assumes that the fishery is responsible for making the case that it meets the MSC standard for a well-managed fishery. The standard is expressed through the principles, criteria and performance indicators in a way that enables the performance of a fishery to be audited by a small expert team of assessors. The fishery has the main responsibility for providing the data and information that proves the case. The MSC process will not normally assume that the case is proven when there is no data or reliable knowledge (the 'objective evidence') that can be used by the assessment team to determine the impacts of the fishery. The lack of data will not be used as evidence that the fishery is not having an unacceptable impact on biodiversity (if nobody is looking for impacts it isn't likely that impacts will be known). In essence, the MSC process places the burden of proof on a fishery to show that it meets the MSC standard. In practice, this means that the fishery should compile all the relevant available data and knowledge on each matter or issue for presentation to the assessment team. Stakeholders also will compile existing data and knowledge for the assessment. Where there is dispute, or where the assessment team considers the evidence is not adequate, the assessment will make cautious decisions, reflected in the scores, in order to ensure that the lack of knowledge is not a basis for awarding MSC certification to a fishery.

This allocation of the burden of proof may permit flippant, malicious or vexatious accusations of impacts that a fishery will need to examine and defend, and perhaps in a costly manner. However, the MSC process discriminates against many such problems by using the expert judgement of the assessment team to determine if such accusations have merit, based on any objective evidence presented, and on experience in other fisheries or in the scientific literature. The intention here is not to prevent matters being raised for consideration in the assessment, but it is necessary to restrict such matters to those that could be of a type or extent that would be determined as important impacts if they were shown to exist in the fishery.

4.9 Resolving issues

A well-managed fishery will either have identified and resolved environmental issues, or be in the process of identifying or resolving such matters. A fishery may be at any of four stages in its approach to environmental issues:

- The fishery may be in denial; this may occur where there is either insufficient data, knowledge or history in the fishery to enable the fishery to make an assessment of the importance of a particular environmental impact. Alternatively, there may be disagreement with stakeholders about the importance of an acknowledged effect of the fishery, such as removal by the fishery of large proportions of an age class of fish that may also be food for important species such as seabirds, marine mammals or endangered turtles.
- The fishery may be in the process of identifying environmental issues that are important, and that should ultimately be addressed. This may be the result of an ongoing and routine research programme in the fishery, or it may be as a result from an ecological risk-assessment process.
- The fishery may have identified the key issues and be proceeding to resolve these issues in a planned and orderly process. This may, for example, involve establishment of research projects in partnership with stakeholders, or with other relevant government agencies or research bodies, in order to secure appropriate data to resolve specific issues. This may include research projects conducted outside the fishery (such as in a fishery that uses similar gear) that have direct relevance to the fishery that is subject to the MSC assessment.
- The fishery may have completed its own internal assessments of environmental

issues, resolved the main issues, and have an ongoing programme of monitoring and research in order to ensure that issues considered resolved are in fact resolved, and to maintain a current awareness of potential new issues.

An MSC assessment will take a very broad overview of progress of a fishery in dealing with environmental issues. First, an assessment would normally consider the context of the performance indicators and if impact issues have been identified, if they are being addressed, and if they have been resolved by the fishery. In addition, an assessment would consider what processes and procedures are in place to achieve these three matters. Each of these are considered important aspects of a well-managed fishery and would normally be the subject of specific performance indicators. Performance of the fishery in resolving these issues is assessed in the context of performance of other fisheries, stakeholder expectations, and practice in other sectors.

4.10   Fishery impacts

Under Principle 2, an MSC assessment considers the impacts of the fishery on the target species, on non-target species caught incidentally in the fishery, on benthic and pelagic habitats, and on species that are ecologically linked to the exploited species. Under Principle 2, the assessment would normally consider aspects of the target species that are not considered within Principles 1 or 3.

4.10.1   *Target species*

Most of the environmental issues associated with the target species will be considered in the context of Principle 1, which focuses on the stock management issues. However, the focus in Principle 1 will normally be on maintaining the populations of the target species for the purposes of exploitation, while in Principle 2 the focus is on maintaining the biodiversity of the exploited species. These are not necessarily the same, and for some species they will translate to be very different objectives. Key aspects of the biodiversity of the target species include the abundance of individuals in space and time, the age-size structure in the populations, their range, and their genetic composition. Each of these should be expressed in a performance indicator in each relevant MSC principle, consistent with the independence of the principles. These latter matters might include under Principle 2, for example, performance indicators such as maintenance of the normal (unfished) size-age spectrum in the population, maintenance of the normal biogeographic range of the species, or the maintenance of the normal range of genetic composition in the species.

Possibly the most contentious of these matters is likely to be the ecological effects of the reduction of biomass that occurs when the fishery withdraws biomass from the population of the target species. Possible effects include direct impacts on species that would normally be prey for the exploited species, but also a large array of indirect ecological impacts that may flow from this. This includes the prospect of 'trophic cascades' where removal of one species by a fishery causes a cascading range of effects in ecologically linked species. Such effects have been found in a range of fishery types and locations, and although these impacts have not always been detected during fishery studies, the prospect of such effects must still be evaluated by a fishery.

The effects of fishery extractions can be assessed from a fishery production perspective, and may be found to be well managed for the purposes of a sustainable fishery but be not sustainable in terms of environmental impacts on dependent species, or on the populations of the exploited species itself. This can occur when a fishery targets populations in a comprehensive and highly efficient way, leaving no refuges, resulting in comprehensive removal of the older age classes. In such cases, the fishery effects on the population of the target species may be detrimental by reducing the genetic composition, eliminating the evolutionary potential of the population, and reducing its fitness to withstand environmental changes imposed by non-fishery factors such as climatic change and extremes. Where target species are dominant species in their local ecosystems, such as high level predators, and large proportions of their populations are removed by the fishery (the usual situation), then their normal prey species may react to the release of predation pressure and this can cause major changes in local ecosystems.

A sustained history of biomass harvest by a fishery provides only limited evidence to assess the ecological impacts of the fishery. While sustained harvests may provide useful information to assess the effectiveness of stock management, it does not provide useful information to assess the broad ecological impacts of the fishery. While a long history of catches may indicate a lack of overfishing (other factors such as climate dynamics or ocean current variability may interact with stocks), a history of sustained catch provides little information about the broad ecological effects of the fishery such as possible effects on species that may grow in abundance in the relative absence of a top-level predator. Also, the type of data often gathered in fishery catch-effort monitoring programmes may not be appropriate for assessing these ecological impacts because they do not normally relate to ecological parameters of the population such as geographic range or the size–age spectrum in the population determined from fishery-independent surveys and monitoring. Similarly, fishery management may be targeted at maintaining a spawning biomass of the target species, rather than, for example, a more ecologically important parameter such as a range of sizes–ages of spawning individuals and locations. This doesn't necessarily mean that populations are not maintained with adequate genetic diversity, but there may be effects that pass unnoticed by the fishery because the monitoring programme tracks other aspects of the population that are most important for stock management and not ecological diversity issues.

Where this knowledge is lacking in a fishery, the most efficient manner for assessing the nature and extent of this potential range of impacts is through the comparative study of fished and unfished areas, preferably fully protected areas where patterns of species composition and abundance might be considered to be representative of the unfished situation. In this case, where differences are detected, the acceptability of such effects will be determined by assessment of the stakeholder acceptability and the ecological importance of such effects. A key aspect of the assessment will be the extent to which any such important changes caused by the fishery are reversible, or are offset by refuge areas that are properly protected and managed so that any fishery-induced changes could be reversible in the absence of the fishery and given a suitable time-frame, say ten years, for an acceptable level of recovery.

4.10.2   Non-target species

The assessment will normally consider the species composition and abundance of all species caught either intentionally or coincidentally in the fishery to determine if the mortality imposed on these species is ecologically important, and subsequently if the fishery is taking adequate steps to reduce this mortality. The nature and extent of the monitoring programme that provides such data and information is also an important aspect of the assessment, since inadequate monitoring systems will provide unreliable data on the catch of non-target species. For some species, such as rare or endangered mammals or birds that may occur in only very low abundances, even a very limited additional mortality imposed by fishing may be of major ecological importance to the populations. Therefore, the design of a monitoring system that can detect accurately even low levels of fishing-imposed mortality can become an important factor in the assessment.

The assessment will, therefore, include the extent to which the fishery has an adequate fishery-independent (or independently validated) monitoring system that will include all relevant species or taxa, and the ecological significance of the mortality that is disclosed by this monitoring system. This could include an assessment of the intensity of fishing in specific habitat types, to determine if there are natural refuges for non-commercial species, the extent of fishing on each habitat type in the region, and the recovery potential of any species where there may be concerns for the population or genetic impacts of the fishery.

4.10.3   Endangered, threatened or protected species

Many fisheries will operate in a region where there are species of very high con-servation value, and some of these may be formally declared under local or national laws as species that must be given special attention. A well-managed fishery will also give special attention to any species that have special stakeholder significance, even if the species is not formally protected by law; such species will be determined by the assessment team taking account of stakeholder submissions, the species conservation status, and any relevant international agreements, conventions, or competent clas-sification system for conservation status such as the IUCN Red List. Species that are protected by law or otherwise are of special significance, are 'icon' or 'charismatic' species, terms used to signify the high level of public and stakeholder interest in managing such species.

The effects of a fishery on icon species will always be a matter of great stakeholder concern, but may also be difficult to assess in detail. Some of the effects of a fishery may be subtle, such as the indirect effect of reducing food availability for an important species, or the effect of a small by-catch of a species that is in very low abundance because of other ecological impacts or natural population stresses. Where an endangered, threatened or protected species is an issue in the fishery, specific details of impacts are required, including the mortality or other effect imposed by the fishery, the population dynamics and status of the species affected, the relative impacts of other fisheries or sources of impact and stress on the populations. Where the fishery is considered to be a major factor affecting the ecological viability or, say, the recovery of a population of an icon species, then the assessment will focus on the fishery effects and on the status of knowledge about the response of the icon species to fishery activities.

Assessment of these impacts is critically dependent on the knowledge base. Data from survey and monitoring programmes may be highly uncertain, or in some cases may not be targeting the most appropriate variables in the context of assessing fishery impacts, and so making a detailed assessment can be difficult. Where data and knowledge are inadequate, the assessment will make cautious decisions and scoring will reflect a precautionary approach to ascribing responsibility for impacts.

The assessment would normally consider all relevant information on endangered, threatened or protected species, and depend primarily on scientifically robust conclusions. The scoring would be based on performance of the fishery in relation to the ecological importance of any impacts, and on stakeholder acceptance of the impacts.

4.10.4 Habitat impacts

The assessment will normally consider the impacts of the fishery on each specific habitat type where it operates. Where there is inadequate data for a judgement to be formed, the assessment team will normally consider the adequacy of any research programmes that may be underway in the fishery or any other sector that may contribute useful information to determining the impacts of the fishery. Also, as for species issues, the assessment will consider the nature and extent of any monitoring programmes that may be operating to maintain a current assessment of habitat issues and conditions.

The types of habitat impacts that may be relevant in the fishery include physical destruction of habitat by trawl gear, by traps, lines or ropes etc, or by lost and discarded fishing gear. The MSC assessment will consider each of these and assess the ecological importance by reference to the nature and distribution of the habitats affected, their conservation status, and their sensitivity to impact or disturbances. For example, a trawl fishery that is exclusively pelagic will have little physical effect on the pelagic habitat (although it may have a substantial effect on the ecology of non-target pelagic species), whereas a demersal fishery targeting the same target species may have much more ecological impact because of the sensitivity of the benthic habitats. However, even in this case, where the fishery leaves refuges, either because grounds are unfishable or are protected from fishing by law, or where habitats being trawled are also widely distributed outside the fishing grounds (and are not under environmental pressure from other fisheries or other sectors), then the fishery effects may prove to be acceptable. Benthic impacts while intensive locally may be of only minor broader ecological significance if the physical structure of the habitat is not overly degraded, the benthic species affected are not locally endemic, are not unique, endangered or threatened, and are found in adequate refuges in the local area implying a reasonable prospect for reversibility of fishery impacts. Determining reversibility of impacts would depend on the nature of the species affected, and on many ecological and oceanographic aspects of the fished ecosystems, the refuges, and the surrounding region.

In this context, it is unlikely that the fished habitats would ever be able to be rehabilitated fully to the condition that existed prior to fishing. This is because first, the habitats themselves will have changed over time even in the absence of fishing, and second, rehabilitation of habitats is a highly vexed question. Benthic habitats that are largely disturbed and then recolonised from distant sources are not likely to fully return to an unfished condition for a very long period of time, and for many habitat types, convergence with unfished conditions perhaps would occur only over centuries, and possibly in geological time. In this sense, fishing of benthic habitats with destructive gear types will often be a matter of stakeholder acceptability rather than a matter of ecological reversibility and rehabilitation which, in most cases, will not be fully achievable on any acceptable time scale.

4.11   Assessment and evaluation

The assessment and evaluation of the fishery is conducted by the team of experts appointed by the certifier, guided by an expert skilled in the process of certification. The assessment team considers the evidence provided by the fishery and the stakeholders both for and against performance against the MSC principles and criteria, and the derived performance indicators (see chapter 5 for more detailed discussion of the standards). The objective of the assessment is first, to identify the issues that are of importance in the fishery, then to evaluate (score) the achievements of the fishery against the performance indicators established for each criterion. The scores for each indicator are aggregated to a final score for each criterion, and the level of this score determines whether a fishery passes or fails each criterion, and subsequently each principle. Each principle is scored independently, and to achieve MSC certification the fishery must achieve at least 80 on all three principles. Where relevant, specific criteria and performance indicators may be applied in more than one principle. This is important to ensure that the assessment considers all relevant aspects of a fishery's performance in determining a score on each principle.

In an MSC assessment, the environmental matters that may be of concern to stakeholders, in addition to the basic question of overfishing, can range very broadly, and may include:

- the removal of the target species may result in a reduction in food availability for dependent species, such as marine mammals;
- introduction of diseases in bait, or the over-exploitation of species in bait fisheries;
- the effect of discarded bait on ecosystems;
- introduction of pests in ballast water, or on the hulls of foreign vessels in the fishery;

- destructive impacts of gear on benthic habitats, or other destructive fishing practices such as pair-trawling (large nets strung between two fishing vessels);
- entanglement of icon species (such as whales) in fishing gear or discarded plastics;
- indiscriminate (highly non-selective) fishing practices, such as non-targeted mesh nets;
- ghost-fishing – the impact of gear lost or discarded from vessels as it continues to catch fish, both target and non-target;
- pollution from fishing vessels or fishing practices.

To determine which matter, from the very large range of possible matters, are important in a fishery, and of concern to stakeholders – including fishers, since some matters such as coastal pollution may also interact with fish stocks – an MSC assessment will normally take a comprehensive and broad approach to stakeholder consultation. As in all areas of an MSC assessment, although issues may be raised by stakeholders, without supporting evidence such matters will not necessarily be accepted by the assessment team as issues to be resolved by the fishery within its management regime.

Generally speaking, the MSC assessment will attempt first to identify the environmental matters that are issues, in the sense that the fishery may have a detrimental effect on specific aspects of the MSC principles and criteria and dependent indicators, then subsequently evaluate the fishery performance against each of the these issues. A well-managed fishery could deal with these issues in different ways depending on the issue and the degree of difficulty in reaching a resolution of the issue.

Environmental issues that face a fishery can fall into one or more of the following classes:

(1) Issues that may be more apparent than real issues – because of poor communication systems this is not acknowledged by stakeholders. Such issues can be routinely resolved by better information and awareness amongst stakeholders of the various aspects of an identified issue. This might include, for example, implementation of an awareness programme for specific fishing activities.

(2) Issues that are more fully the responsibility of an agency or entity not closely connected with the fishery. This might, for example, be an issue of coastal pollution that degrades habitats of importance for a fishery, where the pollution emanates from a watershed that is controlled by an environment protection agency, or some other form of government control.

(3) Issues where knowledge is very limited, and that need detailed knowledge or extensive datasets that currently do not exist in order to reach a resolution. These may include matters such as the ecological effects of greatly reduced natural abundances of the high-level predators that constitute many of the target species of capture fisheries, or the risk of spreading diseases from the use of fresh bait.

(4) Specific and detailed issues where the nature of knowledge and data, or its interpretation, is disputed. This might include, for example, data on by-catch of an important species derived from an observer programme, where the design of

the observer programme is weak and the data thus open to various interpretations.

In each of these circumstances, an MSC assessment will expect the fishery to provide an appropriate and adequate set of procedures for how it will deal with the issues and ultimately reach a resolution. Such resolution will need to be framed within the context of the 5-year life of the MSC certification, but certainly in the more complex issues such as those of class 3 above, such resolution may be staged and conducted incrementally in conjunction with stakeholders, including science agencies, government agencies, local communities and the private sector.

The MSC evaluation would compare the approaches of the fishery to those known from other fisheries, or to the best practices that can be feasible and achievable. The usual evaluation will also compare the fishery practices and achievements with the expectations of stakeholders, and with the capacity of a fishery to address the issue and reach conclusions and specific outcomes that will overcome the issue. Evaluation is generally based on the feasibility and achievability of a set of actions by the fishery that would resolve the issue. Such actions might include a range of response types, including:

- devising and implementing a revised code of practice across the fishery, together with independent data to verify compliance;
- working in partnership with community groups, government agencies (including the fishery management, conservation or science agency), to implement specific research or development programmes of relevance to the issue;
- lobbying government agencies for improved laws, regulations or more vigilant practices in terms of external matters that may affect the fishery. This could include, for example, lobbying for a more transparent and comprehensive stock assessment process that included more precautionary approaches to setting allowable catches (in respect of a specific issue of concern such as over-harvesting of specific age classes). It may also include lobbying for better coastal development guidelines, improved watershed practices, or an improved system of reserves to meet both fishing and conservation objectives for a region.

An MSC assessment would include evaluation of the fishery actions in terms of plans and activities designed to implement these responses. This could include the detailed content of action plans, strategies and procedures designed to implement such responses, the nature and extent of outcomes achieved, and the levels of involvement, acceptance and agreement by stakeholders.

In environmental issues, where there may be a lack of scientific data and much subjectivity in the detail of specific issues, an MSC assessment will normally rely on several lines of evidence in reaching a decision, or a score, against a specific indicator. The lines of evidence will include the judgements of the experts on the assessment team (where they are competent to adjudicate), the judgement of other authoritative bodies where such bodies can provide supportive evidence for their position, data and knowledge from the scientific literature, and the submissions of stakeholders (including the fishery and community groups) but recognising that such submissions must be supported by competent evidence. These lines of evidence would normally be used by the assessment team in relation to all matters upon which it is required to adjudicate.

In situations where there may be a likelihood of environmental risk from the fishery, but there is no data or knowledge of relevance, the MSC assessment would normally seek to examine the extent to which a fishery had taken all reasonable steps to assess that risk. The MSC assessment would normally expect that such potential issues would have been considered by a well-managed fishery at some time in its recent history, and be kept up to date. The process, procedures and outcomes of such consideration and updating processes would be evaluated by the expert assessment team. This would include the nature and extent of the consideration, as well as the veracity of the outcomes. In some fisheries this might be conducted as a formal ecological risk assessment approach in partnership with stakeholders, using a conceptual or quantitative basis depending on the nature and extent of identified hazards.

4.12   Monitoring systems

The possible impacts of a fishery may range over many disciplines of science, all areas of the fishery, and various life stages of both the target and non-target species. A fishery will not be able fully to monitor and evaluate the detail of all possible impacts. However, an MSC assessment would expect to see evidence of a strategic approach to the problem of setting priorities for identifying and responding to environmental issues, and this includes the design and implementation of a monitoring programme that is of suitable scale to ensure that impacts are properly monitored.

The assessment would consider monitoring and assessment programmes in the context of their capacity to detect changes that may be of ecological importance, and including the nature of specific indicators monitored and their relationship to the management plan for the fishery; the time and space scales; the extent of fishery independence; and the synthesis, analysis and reporting procedures. Of specific interest will be monitoring programmes directed towards assessment of major environmental issues, and matters that involve endangered, threatened or otherwise important species. Monitoring systems would also normally be expected to provide summary reports of data and interpretations that are available to the public and interested stakeholders.

4.13   Stakeholder engagement

A well-managed and sustainable fishery is one that has minimal impacts on the environment and ecosystems, species etc, and has a maximal level of acceptability to stakeholders. The MSC assessment process is therefore heavily dependent on stakeholder input to enable the assessment team to be aware of, and assess, stakeholder positions on specific issues in the fishery. Determining stakeholder acceptability of issues is always a matter of ensuring that all the costs and benefits have been exposed and discussed, and that the process for stakeholder consultation has been open, transparent and informs stakeholders accurately so that their opinions may be properly informed.

In addition to making an expert assessment of the environmental impacts of a fishery based on evidence available from within the fishery, the assessment process is heavily based on assessing the nature of stakeholder's concerns to determine if they have substance and are in fact issues that the fishery has failed properly to address. While the fishery and stakeholders may not agree on issues, the substance of disagreement may not be a matter of relevance to the MSC assessment, and normally the details of any issues brought to the assessment by stakeholders need to be fully examined in order to determine their relevance to the assessment and the MSC process.

The existence of a robust process for stakeholder engagement is one of the key characteristics that set it apart from other types of certification. Without a well-engaged stakeholder community, the MSC process for any fishery will not reach a satisfactory set of outcomes, and each assessment team will normally take great care to ensure that stakeholder opinions and comments on environmental issues are both solicited and then carefully considered.

D: Principle 3 – Management Systems

Bruce Phillips

4.14 The guiding principles and criteria

4.14.1 *Principle 3*

The fishery is subject to an effective management system that respects local, national and international laws and standards and incorporates institutional and operational frameworks that require use of the resource to be responsible and sustainable.

4.14.2 *Intent*

The intent of this principle is to ensure that there is an institutional and operational framework for implementing Principles 1 and 2, appropriate to the size and scale of the fishery.

4.14.3 *Criteria*

Management criteria

(1) The fishery shall not be conducted under a controversial unilateral exemption to an international agreement.

(2) The management system shall demonstrate clear long-term objectives consistent with MSC principles and criteria and contain a consultative process that is transparent and involves all interested and affected parties so as to consider all relevant information, including local knowledge. The impact of fishery management decisions on all those who depend on the fishery for their livelihoods, including, but not confined to subsistence, artisanal, and fishing-dependent communities shall be addressed as part of this process.

(3) Management shall be appropriate to the cultural context, scale and intensity of the fishery – reflecting specific objectives, incorporating operational criteria,

containing procedures for implementation and a process for monitoring and evaluating performance and acting on findings.

(4)   The management system will observe the legal and customary rights and long-term interests of people dependent on fishing for food and livelihood, in a manner consistent with ecological sustainability.

(5)   The system will incorporate an appropriate mechanism for the resolution of disputes arising within the system.

(6)   The management system will provide economic and social incentives that contribute to sustainable fishing and shall not operate with subsidies that contribute to unsustainable fishing.

(7)   The system will act in a timely and adaptive fashion on the basis of the best available information using a precautionary approach particularly when dealing with scientific uncertainty.

(8)   The management incorporates a research plan – appropriate to the scale and intensity of the fishery – that addresses the information needs of management and provides for the dissemination of research results to all interested parties in a timely fashion.

(9)   Assessments of the biological status of the resource and impacts of the fishery have been and are periodically conducted.

(10) The management system shall specify measures and strategies that demonstrably control the degree of exploitation of the resource, including, but not limited to:

    (a)   setting catch levels that will maintain the target population and ecological community's high productivity relative to its potential productivity, and account for the non-target species (or size, age, sex) captured and landed in association with, or as a consequence of, fishing for target species;

    (b)   identifying appropriate fishing methods that minimise adverse impacts on habitat, especially in critical or sensitive zones such as spawning and nursery areas;

    (c)   providing for the recovery and rebuilding of depleted fish populations to specified levels within specified time frames;

    (d)   mechanisms in place to limit or close fisheries when designated catch limits are reached;

    (e)   establishing no-take zones where appropriate;

(11) The management system shall contain appropriate procedures for effective compliance, monitoring, control, surveillance and enforcement which ensure that established limits to exploitation are not exceeded and specifies corrective actions to be taken in the event that they are.

Operational criteria

(1)   Fishing operations shall make use of fishing gear and practices designed to avoid the capture of non-target species (and non-target size, age, sex of the target species), minimise mortality of this catch where it cannot be avoided, and reduce discards of what cannot be released alive.

(2) Fishing operations will implement appropriate methods designed to minimise adverse impacts on habitat, especially in critical or sensitive zones such as spawning and nursery areas.

(3) Destructive fishing practices such as fishing with poisons or explosives will not be used.

(4) Operational waste such as lost fishing gear, oil spills, on-board spoilage of catch, etc. will be minimised.

(5) Fishing operations shall be conducted in compliance with the fishery management system and all legal and administrative requirements.

(6) Participants in the fishery shall assist and co-operate with management authorities in the collection of catch, discard, and other information of importance to effective management of the resources and the fishery.

4.15 Approach to the assessment

Of the three principles, Principle 3 is the most comprehensively documented. In general, the questions to be asked to determine the extent of compliance with the criteria are straightforward. However, although there are many questions, they are not ranked in any way, and a number of them seem repetitive or tend to overlap others. This is a problem that assessment teams will tackle when developing the performance indicators and scoring guideposts. There is also a degree of overlap between the questions asked in the three principles, but the viewpoint tends to be different. For example, Principle 2 seeks to determine the effects of the fishery on the habitat. Principle 3 examines whether the fishery uses appropriate methods that minimise adverse impacts on the habitat. In Section 4.7 the point was made that the fishery is responsible for making the case that it meets the MSC standard for a well-managed fishery. This is also the case for Principle 3, and with the current state of development of the assessment processes, this tends to become, for the fishery being assessed, a paper chase for the necessary documents to convince the assessors of the suitability of the fishery for certification.

Only written information tends to be satisfactory as absolute evidence in support of the answers. However the scale of each fishery is different, as is the level of documentation associated with its management. Some very well developed industrial fisheries have fully developed management plans, with objectives, strategic plans to achieve these objectives, and performance indicators to allow assessment of whether the strategic plan is being met, but this is rare. It is unlikely that an artisanal or community based fishery will have such a sophisticated set of documents, although the management arrangements may be well known, adhered to, and produce a satisfactory result.

In other fisheries, even some that are well developed and wealthy, no specific management plan exists for that fishery. This may be for a number of reasons. There may be no plan, or it may be because the principles on which the objectives are based are contained in a higher-level document, which sets the objectives, not for this specific fishery, but for a number of fisheries. As set down in the intent for Principle 3, the form of the information and its degree of specificity is usually 'appropriate to the size and scale of the fishery'.

Not all of the documentary evidence tends to be available in a straightforward form, for a number of reasons. For example, in the Western Australian rock lobster Fishery the management plan for the fishery is a legal document held by the Western Australia Court. It is possible to obtain only an unofficial copy (which is so marked), because at any time the original held by the court might be changed by the legislature. This particular management plan does not include a strategic plan or performance indicators. However, these are built into a separate series of documents, a common situation in many fisheries. It is generally difficult to obtain a single document incorporating all of this information.

4.16   Specific criteria

The authors of the MSC principles and criteria clearly developed them to deal with almost any type of fishery, although their structure suggests that they may have originally been thought of as principally applying to the large industrialised fisheries. It is appropriate, therefore, to comment on some of the criteria and how they may be interpreted.

The fishery shall not be conducted under a controversial unilateral exemption to an international agreement. In general this only applies to large, international fisheries and there is usually ample verbal evidence with regard to the situation. However, depending on the fishery there are usually political overtones that are never stated. In some situations the management plan is ignored or circumvented by the participants, with the acquiescence of the fishery managers. The MSC certification process was set up to attempt to stop, or at least reduce, such practices.

Possess a consultative process that is transparent and involves all interested and affected parties. This is difficult to achieve and varies greatly between fisheries. Increasing consultation and public debate is a worldwide phenomenon, but it is not always seen as satisfactory or inclusive of all parties.

The impact of fishery management decisions on all those who depend on the fishery for their livelihoods, including, but not confined to subsistence, artisanal, and fishing-dependent communities shall be addressed. This tends to occur to a large degree, but it is sometimes difficult to demonstrate that it is undertaken. Only cases where it is not addressed tend to be identified and documented.

Be appropriate to the cultural context, scale and intensity of the fishery. As discussed already, there are vast differences between the systems operating between different sized fisheries. What is really difficult to decide is what level of reduction in the quality of evidence is acceptable with the reducing size of the fishery being assessed.

Observes the legal and customary rights and long term interests of people dependent on fishing for food and livelihood, in a manner consistent with ecological sustainability. Ecological sustainability is the cornerstone of good fisheries management. It is often included in the document of fishery management without a real consideration of what it means. The participants in most fisheries, large and small, have no true idea what it means. The results of the analysis of evidence presented for Principle 1 and 2 must be the critical evidence for this point.

Incorporates an appropriate mechanism for the resolution of disputes. This is essentially a straightforward question. If there are a lot of disputes it is not a well-managed fishery, but in any case where there are few disputes, an effective dispute resolution process must be available.

Shall not operate with subsidies that contribute to unsustainable fishing. Almost all fisheries receive some form of subsidies. Fishers tend not to think of fuel, taxation concessions, unemployment benefits, or even those named as subsidies such as boat building subsidies, as subsidies to their fishery. It usually requires some investigation to identify the type and level of subsidies that are in place. It is usually the level of the subsidies that is important. In some cases they can lead to continued fishing although catch levels are low, and overfishing of the resource may be the result.

Use a precautionary approach particularly when dealing with scientific uncertainty. Clearly this depends on the results of the analysis of Principles 1 and 2. However, in a well-managed fishery, documentation of plans for responses to catch fluctuations should exist, and be precautionary.

Incorporate a research plan. Most fisheries have some form of research being conducted on the target species, and in many cases on the fishery itself. What is needed for success in this assessment is to show that the research is directed at solving the problems of the fishery, not just yielding additional knowledge for its own sake. A prioritised research plan, which addresses industry and sustainability needs, is what is needed, with adequate funding sources to undertake the investigations.

Contains appropriate procedures for effective compliance, monitoring, control, surveillance and enforcement. There are some illegal activities associated with every fishery. This does not mean that they are carried out by the fishers themselves, or with their support. Often the perpetrators are people without the right to fish who illegally harvest the resource. Again, it is the level of illegal fishing that is important. There is no rule of thumb that can be applied but such activities should be clearly identified and documented, as well as the attempts and plans which exist for the reduction in such activities. Illegal catches should be included in the resource estimates, assessed under Principle 1.

The Operational Criteria are relatively straightforward but number 2, the criterion with regard to minimising adverse impacts on the habitat, depends on what is discovered under Principle 2.

4.17   The situation at 2002

The certification of fisheries is a new process and the people involved in fishing, fisheries management and fisheries science are generally unused to the auditing style of approach. By this I mean that a traditional auditor examines a set of books, which are usually self-explanatory, and all the relevant information is available and has been collected and collated prior to the auditor's arrival. The current status of the certification process is that the certifiers often have to advise the client as to what type of information they need to examine, and in some cases have to tell them where they should look to find the information. In addition, interviews with stakeholders often identify additional information that is available, but was not identified by the client.

This situation will change rapidly, and within a few years MSC certification of fisheries will be as commonplace as are the other types of certification (such as ISO) in the community. To a large extent, the experiences of these first MSC certified fisheries will dictate the nature of the certification process to be conducted in following fisheries. Also, the success of the MSC at meeting its objectives will dictate the pace of uptake of the MSC certification by other fisheries. During this time the criteria under each MSC principle are likely to be reviewed and refined, and they may be changed to resolve some of the issues raised in this chapter.

Implementing the MSC Programme Process

Chet Chaffee, Bruce Phillips & Trevor Ward

5

5.1 Introduction

Under the MSC programme the entity that is to be assessed is a fishery – defined in terms of its exploited stocks, fishing effort, locations and timing of extractions, and environmental interactions up until the point of landing of the catch. Any activities aboard a fishing vessel after landing the fish or at shore-side facilities are not within the scope of the MSC fishery evaluation process, unless they have a direct effect on the in-water activities of the fishery. However, all post-landing activities may still be important under the MSC programme if they are significant to tracking the chain of custody of seafood products produced from certified fish. The MSC program for fishery assessments and certification consists of two parts: the MSC pre-assessment and the MSC full assessment and certification. This chapter examines the aspects of both steps in the fishery evaluation and certification process.

5.2 MSC pre-assessment

A pre-assessment is designed to help prospective clients evaluate what it will take to get a specific fishery or fisheries through the MSC evaluation and certification process successfully. This means that the pre-assessment needs to answer several general questions:

- Is there enough information available to properly complete a full MSC evaluation?
- What are the specific responsibilities of the client, the certifier, and other participants in the MSC evaluation process?
- Does the information available at the time of the pre-assessment indicate any specific problems in completing an MSC assessment?
- What would be the estimated costs for a complete MSC certification evaluation, taking account of the range and size of the fishery and any specific problems or issues identified?

5.2.1   Information required for a MSC assessment

To complete an MSC evaluation process successfully, an assessment team must be able properly to identify that the documents and data needed to provide proof that a fishery complies with the MSC standards are available. In addition, the pre-assessment should provide a description of where these documents may be obtained. Both of these pieces of information are important as they will help the client focus its efforts on preparing information for a full-evaluation team should the fishery progress to the stage of the full evaluation.

Although each fishery is somewhat different in size, scope, and complexity, there is a general or basic set of information that is needed by all fisheries. Each fishery that has undergone an assessment has been through the process of identifying and finalising a set of performance indicators by which the fishery is evaluated. These provide the starting point for any new assessment, and as such, provide the foundation for determining what information and data may be necessary in a new evaluation.

5.2.2   Participant responsibilities in a full assessment

It is important to remember that the MSC programme is entered into voluntarily. This means the emphasis is on the fishery providing the evidence that it meets the MSC principles and criteria, and the assessment team providing the evaluation. This separation of rôles is crucial, and it makes the responsibilities of the client and the certifier quite clear and distinct. The pre-assessment helps the client understand these separate rôles, and what the client will be asked to do in a full assessment. The client's responsibility would normally be to submit information, answer questions, provide access to fishery managers, scientists, and others requested for interviews by the assessment team.

The pre-assessment should also point out how an assessment will approach the evaluation process; outlining the steps it will follow (from forming a team through completing a peer-reviewed report) in completing the assessment. The client will be reminded that the experts on the assessment team will not do original research or document retrieval. The assessment team will endeavour to solicit the relevant type of data and knowledge from the client and stakeholders, but is not responsible for any failure to use relevant data or knowledge that could influence the certification outcome but is not provided to the assessment team by the client or a stakeholder.

The pre-assessment should point out to a client the rôles other parties such as stakeholders and the MSC may play throughout the evaluation process. Stakeholders in the fishery, especially those that have established a record of concerns, would be asked to provide input to the evaluation process. The MSC will only become part of the evaluation process if specific information is requested in terms of defining terms or interpreting statements in MSC documents, or if certifiers ask for specific guidance. Otherwise, the MSC is not involved in the specific evaluation process. The rôle of the MSC is limited to overseeing the actions of accredited certifiers to ensure that the MSC procedures are properly applied.

5.2.3   Potential problems in the fishery

During the pre-assessment, the assessor will not only look to see what information is available but will also generally determine the quality of the information and if there are any risks for the client in moving forward to a full assessment. If any significant sources of information show that some aspect of a fishery is clearly in violation of the MSC principles and criteria, or if the information is grossly incomplete, the pre-assessment assessor would indicate the problem and the effect it may have on the fishery in a complete evaluation. One example of this is information requirements for small community-based fisheries. In such fisheries a certifier would try to determine if there was adequate information on the status of stocks. According to the MSC principles and criteria and the MSC Standards Council, a fishery must be able to show that the stock being fished is healthy and not in serious decline or trouble. By 'stock' what is meant is the complete biological stock, not just that portion fished by the community. Therefore, if a community is fishing a small portion of a much larger stock, the MSC still requires that the community be able to provide proof that the entire stock is healthy. While this can often be accomplished through co-operation with government agencies in charge of managing the wider fishery, it is not always easy. If this type of information is available, but lacking in breadth and depth, it may be identified as a problem that needs to be overcome prior to a fishery entering into the full-assessment process.

5.2.4   Estimated costs for a complete MSC evaluation

Perhaps the most difficult, yet most sought after information by any client is the specific cost for completing a full evaluation under the MSC programme. The pre-assessment process requires the certifier to determine the costs of a full assessment. By reviewing what information is available, what interviews will likely be needed, and the size of the stakeholder pool, the set of stock issues to be dealt with, a certifier can better estimate a full cost for a client. This might include for example, the need to conduct interviews with national, state and local fishery management levels in order to fully evaluate how a stock is managed. In addition, a pre-assessment may provide an estimate of timing to complete a project as well as other ancillary information should it be available.

Ultimately, a certification body would prepare a pre-assessment report to the client that contains the following information:

- the fishery management policy objectives, regulations, and practices;
- the fishery's state of preparedness for assessment, in particular, the extent to which the fishery system is based upon the MSC principles and criteria;
- list of stakeholders in the fishery;
- a short description of the fishery;
- general historical background information on the area of the fishery;
- identification of other fisheries in the vicinity but not subject to certification;
- a decision as to whether it will be possible to move from the pre-assessment to the assessment stage;

- a discussion of the key issues and factors identified as potentially troublesome in completing a successful certification assessment against the MSC principles and criteria;
- a budget estimate for conducting a full certification assessment of the fishery.

Until August 2002, all pre-assessments have been considered confidential. The reasoning has been that pre-assessments contain some proprietary information on certification costs. In addition, a pre-assessment report contains the opinion of the certifier, yet the opinions expressed are only based on taking a snapshot of the fishery in a limited time-frame. Although the opinions are meant to help a client evaluate the risks of partaking in a full evaluation process, they may be shown to be inaccurate once a full assessment team completes a thorough evaluation. Having this information publicly disseminated may unnecessarily harm a fishery or a certifier if it is taken out of context and used improperly. Only a full assessment can truly determine if a fishery complies with the MSC standards.

5.3   MSC full assessment and certification

A full fishery assessment under the MSC programme consists of a number of steps. The general idea is for the certifier of record to ensure that the evaluation team is fully qualified, makes transparent the specific sets of information that will be used to evaluate the fishery, provides opportunity for input by stakeholders, evaluates the fishery objectively, drafts a report detailing the findings and the necessary supporting documentation, provides for peer reviews of the draft report, and makes recommendations to the MSC on whether the fishery should be certified. The general outline of the certification process as documented by the MSC can be seen in Table 5.1. Readers should be aware that the MSC certification methodology is currently being updated, and specific aspects may change in time; here we present a broad overview. Current methodologies are available at *www.msc.org*.

5.3.1   *Announce the fishery is starting the MSC process*

The requirement under the MSC programme is that the certifier of record provide public notice of the intent of a fishery to go through the full MSC evaluation process. Specifically, the MSC methodology states: 'Certifier announces ongoing certification on their website' and 'MSC announces ongoing certification on their website'. The intent of these actions is to make sure any interested party is notified that the fishery is in the process of being evaluated under the MSC programme so that interested parties can provide input through the stakeholder consultation process.

Some certifiers have also taken extra steps to ensure that stakeholders are aware that an MSC evaluation is taking place. These steps have ranged from publishing notices in industry publications to publishing notices on lists servers specifically geared to the fishing, management, and stakeholder communities. For example, the

Table 5.1   The MSC certification process. (Adapted from *www.msc.org*)

Step	Certifier Actions	Certification Process	MSC Actions
1		Following a successful pre-assessment of a fishery, a certification contract is signed between the certifier and fishery client.	
2	Certifier announces on-going certification on their website.		MSC announces on-going certification on their website.
3		Certifier's team meet to review pre-assessment; produce draft performance indicators and scoring guidelines; arrange visits to fishery and stakeholder meeting; advertise the certification and stakeholder meetings.	
4	Certifier publishes draft performance indicators, scoring guidelines and dates for the assessment; advertises assessment and invites stakeholder comment.		MSC publishes draft performance indicators, scoring guidelines and dates of the assessment on their website.
5	One month allowed for stakeholder feedback. All feedback must be documented and supported by scientific testimonial etc.		
6		Following stakeholder feedback, certifier's team discuss possible changes to performance indicators and scoring guidelines. Final certification guidelines published.	
7	Certifier publishes revised performance indicators and scoring guidelines on their website.		MSC publishes revised performance indicators and scoring guidelines on their website.
8		Certifier's first visit to fishery and stakeholder meetings.	
8a		[Subsequent visits to fishery as appropriate.]	

Continued

Table 5.1 *Continued*

Step	Certifier Actions	Certification Process	MSC Actions
9		Certification team reviews data and scores fishery against performance indicators and scoring guidelines.	
10		Certifier produces draft report and submits to client for review.	
11		Certifier amends report and submits to peer review.	
12		Certifier amends and produces final report and public summary.	
13		Certifier's certification committee approves certification.	
14		Certification issued by certifier.	
15	Certifier publishes public report on their website.		MSC publishes public report on their website.
16		Certifier plans ongoing surveillance visit.	
17		Certifier undertakes surveillance visit.	
18		Certifier produces report and public summary report.	

Marine Fisheries Conservation Network (MFCN) has an e-mail list server to which most of the major environmental or conservation organisations subscribe, thereby providing an excellent avenue for disseminating information to this sector. Similarly, *BC Fishnet* is a list server to which many subscribe who are interested or involved in fisheries issues in British Columbia, Canada. In addition to these, certifiers may also direct e-mails, faxes, letters, and phone calls to specific organisations identified as significant stakeholders in the fishery. No matter what the means, it is the intent that is critical, and that is to make sure that any interested parties have full knowledge of, and access to, the certification process. Whether the minimum requirement by the MSC to post these notices on the MSC and certifier website is a sufficient dissemination of information, is a judgment that is left in the hands of the certifier.

5.3.2   Establish an evaluation team

The next step in the process of certification is the formation of an evaluation team, a very important step in the evaluation or assessment process. The composition of the evaluation team directly reflects on the intent of the MSC programme to provide a highly respected and technically competent review of fisheries as well as on the credibility of the results to the wider stakeholder community. Selection of an unqualified evaluation team increases the risk of the assessment outcome being challenged by an opposing team of experts, leading to a long drawn-out process that costs the client more and may result in the fishery failing a certification that it could have passed if a genuinely qualified expert team of assessors had been used.

The MSC certification methodology states: 'Central to the demonstration of competence by the certification body is its ability to put together appropriately competent teams of assessors'. In Section 6.4 of the MSC certification methodology, requirements for assessment teams are specifically defined. The MSC points out that the assessment team must have appropriate expertise in fishery stock assessment, fishery impacts on marine ecosystems, fishery management and operations, field experience, regional expertise, regional credibility, third-party product and management system conformity assessment auditing techniques (ISO14011 and ISO10011), and knowledge of the MSC principles and criteria and MSC certification methodology. More specifically, the document points out that:

- 'The fishery stock assessment expert must be a practicing stock assessment modeller with expertise in relevant stock assessment modelling techniques.
- The ecosystem expert must have appropriate experience in the research and/or management of fishery impacts on marine ecosystems.
- The fishery management expert must be a practicing fishery manager or fishery-management policy analyst with a good understanding of relevant fishery-management systems'.

The responsibility for selecting members of an evaluation team falls to the certifier of record. There are no specific requirements under the MSC for the certifier to consult with other parties, only to ensure that the evaluation team members meet the MSC requirements. In the past, certifiers have varied in the practices used to meet this requirement. In some fisheries, certifiers have chosen the team members without consultation, whereas, in other fisheries where controversies have been significant, certifiers have consulted with a wide variety of stakeholders from industry, management, and conservation groups. In the former case, with little controversy apparent, meeting the expectations of stakeholders can be quite different than in the latter case where there are numerous expectations regarding the transparency, objectivity, and rigour of the MSC assessment process. For example, in the Alaska pollock fishery currently under evaluation at the time of writing, the certifier of record, Scientific Certification Systems Inc., spent six months polling the fishing industry and stakeholders in the conservation community to find evaluation team members that were technically competent, of high standing in their respective disciplines, could maintain objectivity, and provided a balance of viewpoints (also referred to as balanced bias) so that when a consensus was reached on the sustainability of the fishery it would have a higher probability of being accepted by the majority of interested parties.

5.3.3   Develop performance indicators and scoring guideposts from the MSC principles and criteria

Once a team is selected, the next step in the MSC process is to develop specific indicators of performance by which to measure the fishery against the MSC standard. The MSC principles, by necessity, are general statements about the categories of issues that need to be addressed in a sustainable fishery (similar to the FAO *Code of Conduct for Sustainable Fisheries*). The MSC criteria are slightly more specific than the principles, providing a more refined set of concerns within each of the principles.

To accomplish the task of a performance evaluation, even more specific questions need to be developed under each MSC criterion such that an evaluation team has a list of the components (known as performance indicators in the MSC system) that define sustainable fishery management as provided by the MSC standard. In addition, the evaluation team must provide the written basis it will use to determine if each component is performing properly. In the MSC system of evaluation the written bases for scoring are referred to as the 'scoring guideposts'. The scoring guideposts outline what constitutes theoretical best practice (100 guidepost), practices sufficient to maintain a well-managed fishery and be in compliance with the MSC standard (80 guidepost), and practice that is minimally indicated for the fishery to be considered for certification (60 guidepost).

The draft performance indicators and scoring guideposts must then be submitted for public comment for a period of no less than 30 days. After the public comment period is closed, the certifier reviews the comments and revises the performance indicators and scoring guideposts and posts them on the MSC website as well as the certifier's website for public access by any and all interested stakeholders. To avoid each new assessment starting from scratch, the MSC certification methodology requires each assessment team to begin the process of developing the performance indicators and scoring guideposts by using those already developed in other fisheries. The intention here is to (a) help the certifiers perform the job in an efficient and cost-effective manner, and (b) hopefully to build in some internal consistency between fishery assessments.

Some have also suggested that the MSC could build greater internal consistency between fishery assessments by developing an official codified set of performance indicators and scoring guideposts that stand as the starting point for every fishery assessment. This official set might be developed through a workshop with fishery scientists and managers from around the world, much as the original MSC principles and criteria were developed. The draft document could then be sent out for peer and public review and revised as necessary from the comments received. This would not be unlike the processes in many fisheries where management plans or other regulatory changes are drafted and then set forth for public review and comment before being finalised. This set of performance indicators and scoring guideposts could then be reviewed by assessment teams at the start of each new fishery assessment and modified as necessary to accommodate the size, scale, and complexity of the fishery under review. Certifiers would then provide a written statement of proposed changes to the MSC for approval before using them in the fishery evaluation making the process more formal and improving consistency between assessments.

5.3.4 Compiling and reviewing information

Once the draft is finalised for the performance indicators and scoring guideposts, the certifier of record can begin the process of compiling and evaluating information on the fishery or fisheries under investigation. According to Section 9.1 in the MSC certification methodology, the certifier and the assessment team are responsible for the compilation and analysis of all available and pertinent information. The MSC methodology does not detail any specific operational details on how this information is to be collected, verified, and reviewed. The only requirement is that the assessment team allows information to be input to the process from any and all interested stakeholders. This helps ensure that the information compiled is not biased in any given direction, but covers the variety of interests. There are three aspects to the process of compiling information before the fishery is evaluated; a description of each follows.

Accumulating information from the fishery, fishery managers and scientists

Leaving certifiers to collect their own information has implications for the cost and integrity of the certification process. Even an expert team of scientists that are knowledgeable in the required disciplines (stock assessment, ecosystem impacts from fishing, and fishery management) and in the fishery or fisheries under evaluation, would take a long time to collect all relevant information. There is also a high probability that the team would be unaware of a number of studies, reports, databases, and other documents that may be pertinent to verifying current management practices and scientific aspects of the fishery. Even more likely, an assessment team would not have access to pertinent memos, documents, budgetary data and other sources of information internal to the management and research agencies.

To overcome the excessive time requirements, and to improve the prospect of accumulating the breadth and depth of information necessary to evaluate a fishery fully, certifiers require the client to submit documentation that proves the fishery meets the specified performance indicators (i.e. the MSC principles and criteria). This tactic ensures that the evaluation team will get a substantial set of information from the fishery under evaluation and is able to evaluate fairly the fishery situation.

Accumulating information from stakeholders – fishing organisations, conservation groups

The MSC requires that the certifier place announcements in whatever media (newspaper, industry publications, web-based announcements, electronic lists, etc.) make the most sense for disseminating a request for any and all interested stakeholders in the fishery to provide input to the evaluation team. This step allows the certifier to hear different points of view, which can be very useful in understanding two things: what problems may exist in the fishery, and what opinions must be taken into account to bring credibility to the outcome of the certification.

Reviewing and understanding accumulated information

Once the team has had a chance to request and receive information, the typical next step is to meet directly with whichever parties may be necessary to clarify points, get additional information, or better understand procedural issues in the fishery. This typically means meeting (face to face or by tele-conference) with fishery managers, research scientists, and stakeholder groups that have shown interest in the process. More than one meeting may occur with any one group depending on the size and complexity of a fishery or fisheries. Highly controversial or large-scale fisheries typically require more than one meeting with each stakeholder group, fishery managers, and scientists before the assessment team determines that it has complete enough knowledge on the wide spectrum of issues and data to complete the assessment.

5.3.5   *Evaluating and scoring the fishery*

The MSC methodology for the final evaluation of fisheries requires that an assessment team scores a fishery against the set of performance indicators developed by the team. To facilitate the scoring of the fishery, each assessment team uses analytic hierarchy process (AHP), a decision support tool, to assist the assessment team in prioritising, weighting, and scoring the performance indicators. AHP provides a formal structure for assessment teams to use (Williams, 2002) in considering how well a fishery performs against the developed set of performance measures. Decision support tools are used extensively in many other areas of management decision-making and were chosen by the MSC to improve consistency between assessments by providing a structure for team members to discuss their views and come to a consensus decision regarding a fishery's overall performance against the MSC principles and criteria.

In general, there are three steps to the process of scoring a fishery using AHP: establishing the performance indicators and their hierarchy; prioritising and weighting the performance indicators; scoring the performance indicators. At the beginning of the assessment process a hierarchy of performance indicators (known as indicator sub-criteria in AHP) and operational sub-criteria are developed by the assessment team under each MSC criterion. Performance indicators are measurable factors that are scored by the assessment team, while operational sub-criteria are factors that require additional explanation through subsets of performance measures.

Weighting occurs at each level of the hierarchy independent of other levels of the hierarchy (see Fig. 5.1), and within groupings at each level, with weights assigned to all factors (both indicators and sub-criteria). Weighting (the act of assigning values

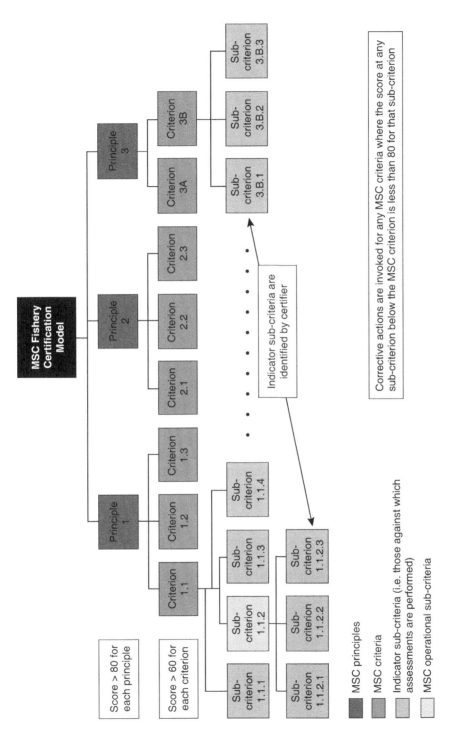

Fig. 5.1 The MSC-certification hierarchy (from Williams, 2002; MSC, 2002).

between 0 and 1) is accomplished in AHP using pairwise comparison (MSC, 2002) where each indicator is compared by the team to every other indicator within the same level of the hierarchy, creating a matrix of comparisons which can be utilised to develop the final prioritised weightings. The purpose of this step is to allow each fishery to be scored against the same or similar sets of performance measures, while acknowledging that different sets of information and data will have different priorities and importance in different fisheries. In this way, the MSC process can be utilised equally in all types of fisheries yet be flexible enough to allow the assessors to identify the priority of the factors.

Once weighted, a separate step is taken where each performance indicator is scored on a scale of 1 to 100 indicating how well the fishery performed for that given indicator. This is accomplished, usually, through a group discussion where the assessment team comes to a consensus decision on the score to be assigned to each indicator using the scoring guideposts. Each assessment team member must participate in the review discussion and provide a score that he or she can support through citation of specific data and information. If significant differences occur between team members, then further discussions take place to identify the cause of the differences and determine what actions need to be taken by the team to come to a consensus. If a consensus cannot be reached, this is an indication that additional information is needed by the team to resolve the problem. Once consensus is reached, a score is assigned to the indicator.

AHP can also be utilised by allowing each assessment team member to score individually and then average the scores, however, this is not the preferred course of action as it does not allow for the identification of differences or the resulting discussions to determine the cause. It is preferred, and required by the MSC certification process, that assessment team members openly discuss their opinions, identify their differences, and then work to come to a consensus score by thoroughly discussing the underlying issues. This helps ensure that the results are not biased by one opinion or one score and that all team members are satisfied that the issue has been properly addressed and can be fully substantiated in the final report. Ultimately, each indicator score is multiplied by its weight to obtain a weighted score for the indicator. All weighted scores for the performance indicators at each level are summed to provide normalised scores for the levels above (the operational sub-criteria, MSC criteria, or MSC principle).

The MSC system is set up such that the normalised performance scale is arbitrarily chosen to be from 0 to 100. For a fishery to pass and be allowed the opportunity to carry the MSC eco-label, it must receive normalised scores of 80 or greater for each MSC principle. However, individual performance indicators may achieve (raw, unweighted) scores of less than 80, and the fishery may still pass certification. If indicators score below 80, this identifies an area in the fishery where a deficiency (termed a non-conformance in the ISO systems) exists and must be corrected. The certifier must identify the deficiency and propose a solution to bring the score to the 80-level benchmark. The solutions under the MSC methodology are known as conditions (corrective actions under ISO), these spell out not only the required solution but the timeframe allowed to make the improvements. Should a fishery pass 80 for all three

MSC principles, yet have several indicators that score below 80, it can be awarded a conditional certification if the client agrees in writing to meet the conditions set forth by the certifier to remedy the deficiencies in the timeframes allotted. If any indicator scores below 60, the fishery may not be certified until after the problems are fixed, even if the overall normalised scores for each principle are above 80.

According to the MSC methods, certifiers are to work interactively with clients to establish reasonable and achievable conditions in a fishery. The specific language in the MSC Certification Methodology reads:

'The efficacy of the condition-setting process is enhanced when the client is in agreement that the specified actions are an appropriate response to the identified deficiencies. Accordingly, the assessment team is encouraged to arrive at conditions through open dialogue, with the fishery receiving certification.'

5.3.6 Draft evaluation report and submit to client for comment and clarification

The fundamental purpose of the assessment report is to explain to interested parties how the assessment team came to its recommendations regarding the proposed certification of a fishery. This means explaining the information base that was reviewed by the assessment team, the strengths and weaknesses found, and the basis for assigning the scores (the specific MSC reporting format is shown below). Where deficiencies in the performance of the fishery against the indicators are found, it is also the certifier's job to spell out the deficiencies and the conditions for correcting the deficiencies.

The original MSC methodology required that two reports were written; a full report for submission to the client and a public summary report for dissemination to the public. This was done for the first five fishery projects that were completed under the MSC programme (Western Australia rock lobster, Thames herring, Alaska salmon, Burry Inlet cockles, and New Zealand hoki). However, this requirement came under attack and led to formal disputes by stakeholders interested in seeing all the details and not just receiving a summary. As a result, the MSC changed this requirement when it modified its dispute resolution policy such that full reports are now to be posted for public review and comment prior to making any final determinations about the certification of a fishery.

Once completed, the draft report is first circulated to the client for internal review. This is the last check by the client to determine if the assessment team has missed any specific details, misunderstood data, or generally made factual errors. The main components of the main assessment report are summarised in Box 5.1.

Peer review of draft report

The draft report must be peer reviewed by competent experts with expertise similar to those of the assessment team; therefore, the certifier must use the same qualifications in choosing at least two peer reviewers. According to the MSC dispute resolution

Box 5.1   A summary of the MSC main assessment report components (modified after the MSC Certification Methodology).

1. **Summary**
   - The summary shall provide a brief description of the assessors, the process, the events that occurred, the main strengths and weaknesses of the client's operation, the recommendation reached, and any conditions attached to the certification and the time-scale for compliance.
2. **Background to the Report**
   - Names, qualifications and affiliations of the assessment team and peer reviewers;
   - Summary of any previous assessments of the client's operation, conclusions reached and past compliance with specified conditions;
   - Itinerary of field activities, description of main activities and locals inspected, and names of individuals contacted during field inspections;
   - People interviewed and a summary of information obtained.
3. **The Fishery Management Operation**
   - A concise summary of basic information as to the management operation (e.g. any ownership, history and organisational structure), the sea area that was evaluated, species types, management history, fishing practices, historical fishing levels, other resource attributes and constraints, user rights (both legal and customary), the legal/administrative status of the operation and involvement of other entities including responsible government agencies.
   - It is recognised that some fisheries will have formally documented management plans whilst others will have an informal arrangement of documentation. This shall be appended as an annex to the certification report. It shall contain information on:
     - area of operation and jurisdiction;
     - history of fishing and management;
     - details of decision-making processes, including the recognised participants;
     - objectives for the fishery;
     - outline of the fishery resources including particulars of life histories;
     - outline of fleet types or fishing categories participating in the fishery;
     - outline of status of the stocks as indicated by stock assessments;
     - description of the aquatic ecosystem, its status and particular sensitivities;
     - details of non-fishery users or activities, which could affect the fishery;
     - details of those individuals or groups granted rights of access to the fishery;
     - description of the measures agreed upon for the regulation of fishing;
     - specific constraints, e.g. details of any undesirable by-catch species, their conservation status and measures taken to reduce this as appropriate;
     - particulars of arrangements and responsibilities for monitoring, control and surveillance and enforcement.

Continued

Box 5.1   *Continued*

4. **Evaluation Procedure**
- The report shall describe the MSC principles and criteria, the scoring guide-lines associated with each criterion, and the weights of relative importance assigned to each criterion.
- The report shall describe the methodology used, including sample-based means of acquiring a working knowledge of the management operation and sea base; describe the scoring process (e.g., group consensus process) and the decisions rule for reaching the final recommendation (e.g. aggregate criterion-level scores must all exceed 80).

5. **Evaluation Results**
- The report shall enumerate and justify the scores assigned to each of the criteria.
- The report shall summarise the main strengths and weaknesses encountered; distinguish recommendations from conditions. The balance of strengths and weaknesses should be consistent with the final recommendation as to certification.

6. **Formal Conclusion and Agreement**
There shall be an appendix to the team's report that shall include a formal statement as to the certification action taken by the certification body's officials in response to the recommendation. Additionally, appended to the report shall be a copy of final contractual agreement reached between the certification body and the client operation and, a copy of the certificate (if awarded).

policy, the peer review must happen prior to a final determination being made by the certification body on the certification of a fishery. The peer-reviewer comments must be appended to the full report, and the assessment team, in conjunction with the certifier, must present evidence of modifying the report based on the peer-review comments, or present reasoned arguments for why the comments did not produce any changes.

Public comment period for draft report

Under the policies of the MSC (adopted in 2002), the final report is now to be posted for public review for 30 days prior to any determination being made by the certification body about certification of the fishery. This allows stakeholders to comment on the entire process prior to a final determination being made rather than after the final determination, which was designed to help prevent potential disputes after certification was issued and a label applied to products. This process was adopted in response to complaints about the transparency of the system and the inability of stakeholder groups in the conservation community to comment fully on the factual basis of certifications before they were finalised.

Determination by certification body

Once all processes are complete (peer review and public comment), the final report, along with the peer review and public comments, is handed to an independent panel within the certification body (also known as the accredited certification company). The certification panel makes the final decision with regard to the certification body's determination. This determination, along with the final report is then passed to the MSC.

Dispute resolution

The MSC has adopted a dispute resolution policy (see Chapter 6) that requires public access to the draft report for 30 days prior to the final determination by the certification body. If, after this, an interested party decides to file a dispute over the findings and recommended determination, there is a specific process that is required. First, a dispute can only be filed by parties that have materially participated in the assessment process. Next, an official form must be filed with the MSC notifying the MSC, the certifier, and the client of the intent to file a dispute. After filing the notification, the dispute report must be filed with the certifier. The certifier then has a specified number of days to respond to the dispute either upholding its initial determination, or modifying it as deemed necessary by the certifier. If the disputing party is not satisfied with the certifier's response, the dispute can be re-filed directly with the MSC. At this juncture, the MSC forms a dispute panel of three or more members with no interest in the fishery or its associated products. The panel reviews all arguments against the determination either regarding the process used or the content of the report. The MSC panel's decision is final and no further disputes can be lodged.

5.3.7 *Certification*

Once a fishery passes all phases of the process, including the dispute process, a certification can be awarded by the certifier. A certificate is issued that states the fishery is certified, and specifies the starting and ending dates of the five-year certification period.

Certification monitoring

Once a fishery is certified, it is required that at a minimum, an annual monitoring be performed by an accredited MSC certifier. This is typically the original certifier, but that can be changed by the certified fishery. The annual monitoring assures that the claim in the marketplace of a sustainable fishery is being maintained – that is that the fishery is maintaining compliance with the MSC principles and criteria. To accomplish this task, the certifier must review several things:

- compliance with any stated conditions in the certification report;
- on a random basis, select areas to inspect within the fishery to determine if the fishery is still compliant with the MSC principles and criteria;
- the views of managers, scientists, industry and stakeholders;
- potential changes in management structure;
- changes, additions or deletions to regulations;
- personnel changes in science, management or industry to evaluate impact on the management of the fishery;
- potential changes to scientific base of information.

Monitoring, also known as surveillance visits, can also happen more often than once a year if the certifier finds cause to be concerned over specific activities in the fishery. Cause can range from non-compliance with stated conditions of certification to impending changes in fishery regulations. At a minimum, each certified fishery will be subject to four annual surveillance audits during the 5-year certification period. At the end of this period, the fishery must once again go through a full assessment to renew its certification.

Dispute resolution and the MSC

Duncan Leadbitter & Trevor Ward

6

6.1 Introduction

The decision-making processes used by the MSC (discussed in Chapter 3) are based on a scientific assessment of conditions in a fishery but also involve consultation with a range of interests, including those who may potentially be opposed to the certification of a fishery. Providing an avenue for evaluating the concerns of those dissatisfied with the results of any decision-making process is a fundamental component of modern, natural resource management. In the absence of any judicial or government mediated dispute mechanism for such a global, independent programme, the MSC needed to establish a process that provided an unbiased and trustworthy avenue for those with grievances to have them heard and evaluated, and to be assured that appropriate responses can be activated. This is stated by the MSC as follows:

'To increase the public accountability of the certification process stakeholders shall have access to dispute resolution procedures which ensure that disputes, grievances, complaints, and appeals are dealt with by the certification body in an equitable manner.' (MSC Accreditation Manual – *www.msc.org*).

6.2 Historical background

The architects of the MSC fishery-assessment system made provision for a grievance process when the accreditation manual was created in 1996–97. It was incumbent on the certifier (see Chapter 5.3) to make provisions for a complaint as follows:

'The certification body shall specify and implement policy and procedures for addressing disputes, grievances, complaints or appeals which are related to the accredited certification programme.' (MSC Accreditation Manual – *www.msc.org*).

The certifier was expected to handle any dispute, at least in the first instance. However, if there was no agreement amongst the parties then the dispute could be referred to the MSC:

'The dispute resolution procedures of the certification body shall have been fully implemented before any dispute is referred to MSC.' (MSC Accreditation Manual – *www.msc.org*).

However, the detail of the process was incomplete and it was the lodgement of the appeal against the certification of the New Zealand hoki fishery that provided the impetus for details to be devised.

By mid 2001 it was apparent that: there was no time limit as to when a dispute could be lodged; there was no procedure for how the MSC should handle any dispute; there was no guidance as to the scope of any dispute that should be entertained by the MSC; there was no timetable for the processing of any dispute. Moreover, there were legitimate concerns by some groups that a certified fishery could be marketing product in the market place when there was legitimate concern as to whether it had met the MSC standard. Proposals for addressing these issues were discussed within the MSC in early 2001 and, in the absence of a stakeholder council or technical advisory board (both of which were to be created as a result of a governance review the results of which were agreed by the board of trustees in mid 2001), discussions were held with a small number of stakeholder groups prior to advice being put to the board of trustees.

The new dispute procedure developed addressed the issues raised above. One consequence was the closure of opportunities for disputes against the Alaskan salmon and Western Australia rock lobster fisheries. As this change occurred midway through the early stages of the hoki dispute, the board of trustees made an exemption for this certification (see below). However, it was the decision to allow a dispute to be heard before any decision on a certificate was made that generated concern by a certification company and its client, which had already entered the full assessment process. The basis for this concern was a belief that the new approach could usurp the certifier, as an independent decision maker, and transfer the decision-making powers to the board of trustees. A number of US-based environment groups were concerned about the imposition of a dispute-lodgement fee, and that the avenue of appeal was restricted to the five-yearly certification assessment and there was no avenue of appeal against the annual audits.

The MSC entered into a period of negotiation with the concerned parties and a number of changes to the process were made. This dispute procedure, known as the interim objections procedure, applies to certification assessments that were commissioned prior to October 2001. In July 2002, the MSC Board of Trustees agreed to some minor modifications to the interim procedure, thus it has now become the procedure to be used for all subsequent certification assessments. The results of the hoki dispute, however, may be used to inform further changes to this important aspect of the MSC certification system. The recommendations by the independent objections panel with regard to any policy aspects of the MSC Certification

Methodology, including the process for resolving disputes, will be submitted to both the MSC board of trustees, its technical advisory board and stakeholder council.

6.3   The dispute procedure

Bearing in mind the fine tuning which may occur in the future, the fundamental aspects of the dispute procedure will remain as follows:

(1)   The first avenue of recourse for an aggrieved party is the certifier. Upon receipt of any complaint the certifier is required to review the determination made and discuss this with the complainant. If agreement cannot be reached the aggrieved party can take their concerns to the MSC board of trustees.
(2)   Upon the receipt of any complaint the board of the MSC must establish a dispute panel chaired by a board member with no interest in the fishery. To advise the panel there should be a minimum of two eminent scientists, one of whom must have had experience on either an MSC certification team or as a peer reviewer of an MSC certification.

The dispute procedure allows for both a merit dispute (i.e. a dispute over the certification determination) and a procedural dispute (i.e. a dispute over whether the certifier properly followed procedures).

After hearing submissions the panel may allow a certifier's determination to stand or require the certification body to investigate in more detail the issues of concern and submit their investigations to the panel. After a period of consultation between the panel and the certification body, the panel is empowered by the board to make a decision about whether or not the fishery in question should be certified. The decision of the dispute panel is final. Once this decision has been made, if the fishery is to be certified, a certificate can be issued and no further appeal will be permitted until the fishery is re-assessed in five years time. If a certificate is not issued then the applicant has the opportunity to address the issues of concern and reapply in due course.

6.4   The New Zealand hoki dispute

The certification of the New Zealand hoki fishery was disputed by a New Zealand-based conservation group, the Royal Forest and Bird Protection Society (RFBPS). In April 2001 it lodged a dispute with the certifier, Société Générale de Surveillance (SGS), raising a large number of issues with not only the operation and impacts of the fishery but the conduct of the certification assessment itself. Some of the range of concerns raised by the RFBPS include the following:

- the indicators and scoring guideposts used to assess the fishery were inadequate;
- the approach to, and amount of time for stakeholder consultation was inadequate;
- one of the stocks of hoki was at risk of overfishing and that measures to address this were not in place;

- the fishery had a known impact on New Zealand fur seals and probably had a greater impact on some threatened seabirds than was known;
- the fishery operated in a climate of information scarcity in regards to the marine environment in which it operated;
- some of the management measures used to control the fishery and ensure compliance were inadequate.

After deliberation, SGS reaffirmed its decision to certify the fishery but did note that subsequent to the period in which the original certification assessment had been made there had been some changes in some aspects of the fishery which required attention, an example being the listing of several species of seabirds as 'threatened' according to the International Union for the Conservation of Nature (IUCN).

In December 2001 the MSC board of trustees wrote to the RFBPS seeking their intentions as regards an appeal to the board as it wished to adopt fully the new dispute procedure that put a time limit on the lodgement of disputes. The RFBPS chose to re-lodge the same dispute that had been submitted to the certification body. The board of trustees appointed a dispute panel chairperson at its January 2002 meeting. The terms of reference for the panel are:

(1) To evaluate the scoring guidelines and comment on:
  (a) whether the range of indicators is sufficient to capture an assessment of all aspects of the fishery relevant to the MSC principles and criteria;
  (b) whether the scoring guideposts were set at a level sufficient to identify a fishery that meets the MSC principles and criteria as being sustainable and well-managed.
(2) To evaluate, on the basis of the information available to the certifiers, the various claims put forward by RFBPS, in the context of the judgements made by the assessment team against the MSC principles and criteria.
(3) To comment on whether the corrective action requests are sufficient to address the identified inadequacies and so result in a fishery that would meet the MSC principles and criteria for a sustainable and well-managed fishery.
(4) To determine whether progress to date and the proposed action plan improve the status of the fishery with respect to the MSC principles and criteria; and whether together they would result in a fishery that would meet the MSC principles and criteria for a sustainable and well-managed fishery, using as information the following reports:
  (a) the SGS audit report dated 23 January 2002, which documents action taken by the Hoki Fishery Management Company to address the corrective action requests;
  (b) the draft corrective action plan.
(5) Provide any comments on areas where the MSC assessment process could be improved.

Two eminent scientists were appointed to conduct a review of the scientific aspects of the dispute and a retired judge was appointed to provide a synthesis and a recommended course of action. The panel was chaired by a member of the MSC board of trustees who has no financial or other interests in the hoki fishery. The review was conducted systematically following the structure of the reporting on each indicator in the certification report. The first step was to examine the score (minimum or pass) awarded to the fishery during the initial assessment, accompanied by the scoring guideposts used for the indicators and the main factors taken into account in evaluating the fishery against the guideposts. If a corrective action was raised by the certification body, its nature also was noted. Next, the nature of the complaint on the indicator was summarised, as were the actions recommended by the RFBPS, if any.

Following those summaries, the dispute panel described the actions that were taken by the fishery between the initial assessment and complaint and the time of the review, whether in direct response to a corrective action request or otherwise initiated. The panel was particularly informed by the corrective action plan of 13 May 2002 of the Hoki Fishery Management Company, the draft conservation services plan 2002–3 of the New Zealand Department of Conservation, and the proposed fishery research services for 2002–3 of the New Zealand Ministry of Fisheries. These summaries helped the panel to understand the issues under dispute, both from the time of the initial assessment and the time of the complaint, as well as any actions that had been taken or are planned for the future.

Having gained a clearer understanding of the background to the dispute and the current status of the fishery, the next step in the review was to evaluate the scoring guideposts used in the initial assessment by the certification body. This evaluation was based upon the dispute panel's interpretation of MSC expectations for the pass or fail guideposts on the indicator. The guideposts used by the certification body may be considered to interpret the MSC standards appropriately, too stringently, or too generously. In order to evaluate the consistency of the application of the guideposts within the framework determined by the certification body, the scoring of each indicator was assessed by the disputes panel. This evaluation by the panel could potentially have concluded an assessment was consistent with the certification body's own guideposts, or was inconsistent by being either overly strict or overly generous.

Next, a summary was prepared of the key aspects of the current status of the hoki fishery, management, research, monitoring and planning, relative to each indicator and the issues of contention between the initial assessment and the complaint. And finally, the reviewers presented and explained their evaluation of the appropriate score for the hoki fishery relative to its current status on the indicator. This evaluation took account of the developments since the initial assessment, and any modifications to the SGS guideposts that they considered appropriate to capture the intent of the MSC standard more closely.

The final result of this dispute has not been determined as this chapter goes to print. However, irrespective of the outcome there is little doubt that not only have the procedures of the MSC improved as a result but, whilst the dispute itself was being addressed, the fishery was moving to address many of the issues raised by the complaint and the certifier.

6.5 Conclusions

The objections procedure, described above, to some extent reflects the problems of timing and the lack of a fully developed objections procedure that could be applied to the hoki assessment and certification. Now that a more transparent and effective objections procedure has been developed, possible objections to MSC certifications in the future should be easier to resolve, and should take much less time than the hoki dispute. The existence of the MSC objections procedure and the history of application are expected to act as a stimulus for all parties to ensure that all relevant matters are addressed and resolved in the certification process. It also assists to reduce the uncertainty about how the certification assessment process is applied, the nature of evidence that is acceptable, and about the standards of stakeholder consultation that are expected.

Over time, as more fisheries go through the certification process, policies and procedures relating to the MSC certification methodology including objections, will evolve in a structured and strategic way. This evolution will be based upon lessons learned, evaluation and analysis of the application the MSC standard and relevant new information from the scientific and policy-making communities. It will be facilitated through the MSC board of trustees and its technical advisory board and stakeholder council. The need for adaptation or change will be balanced against the need for stability and predictability for certification bodies and their fishery clients who operate in a business environment.

MSC Chain-of-custody Certification

Peter Scott

7

7.1 Introduction

The MSC fisheries-certification programme consists of two separate and distinct components – fishery certification and chain-of-custody certification. Fishery certification (as described in earlier chapters) is performed against the MSC fishery standard – the MSC *Principles and Criteria for Sustainable Fishing* and a set of performance scoring guidelines unique to that fishery. In an analogous manner, a MSC accredited certifier performs chain-of-custody certification against the MSC chain-of-custody certification standard. The chain-of-custody certification has two distinct purposes: to guarantee the provenance of a pack of processed or unprocessed MSC certified fish. This means that a pack bearing the MSC logo does, in fact, contain fish that emanates through an unbroken chain from a MSC certified fishery; and is a mechanism for further income generation for the MSC.

The MSC is a non-governmental organisation, and a registered charity, that relies on income from altruistic foundations or use of its logo through licensing agreements on certified fish products. Every new certified fishery provides a further increment to the financial security of the MSC through the income from the licensing of the MSC logo used on those fish products. To overcome the conflict of the 'for profit' nature of logo licensing with the non-profit and tax-exempt status of the MSC, it has established a wholly owned 'for profit' company to deal with MSC logo licensing (see Chapter 3.14). The use of the MSC logo on fishery products is only permitted where there has been independent verification that the product originates from an MSC certified fishery. The chain-of-custody certification provides this verification.

7.2 Who is the typical MSC chain-of-custody certification client?

Any person or organisation wanting to apply the MSC logo to a fish product must first obtain a chain-of-custody certificate. In most cases, this will be some establishment that is selling fish products directly to the consumer such as a retailer (typically supermarkets) or a restaurant chain. However, other businesses may also use the logo to promote their support and use of sustainable seafood. For example, wholesalers, distributors, and value-added processors may wish to use the MSC logo on finished product being sold to retail or food service to help promote their specific brands. Chain-of-custody certification is not required where the MSC logo will not be applied directly to specific products and where no claim with respect to MSC certification is made about products. In a MSC certified fishery with different supply chains, only some supply chains may become certified due to individual business decisions. Products with a certified supply chain will be eligible to carry the MSC logo; products with a non-certified supply chain will not be eligible to carry the logo.

In the case of a multiple-branch or multi-divisional business, a certain amount of self-regulation is permitted by the MSC. Examples would include supermarket chains, restaurant chains, and trade associations. In these cases, a MSC chain-of-custody certificate would be sought by the head office or divisional office. The conditions associated with the issue of that chain-of-custody certificate would require the business to add the requirements of the MSC chain-of-custody certification standard to its internal audit programme covering the individual entities or members within the business. This is to ensure that each outlet is in compliance with the requirements of that standard. The chosen MSC chain-of-custody certifier will ensure that these audits are effective during their on-site routine-surveillance visits.

Some reticence to use the MSC logo on product has been expressed by some retailers because of a fear that positive statements (by use of the MSC logo) on one fish product might cause the consumer to infer negative statements about other products that do not carry the logo. Although consumer-based research has shown that this is not the case, this fear still lingers. Since the MSC programme is voluntary and the use of the logo is not mandatory, the MSC must find better ways to educate both consumers and businesses about what the MSC logo stands for and how it can be used if the MSC hopes to see its label proliferate in the marketplace.

7.3 Where does chain-of-custody certification take place?

The MSC fishery standard assesses the traceability of the fish from a certified fishery up to the point where that fish leaves the fishing vessel. The MSC chain-of-custody certifier must consider all parts of the supply chain, from fishing vessel to end consumer, (Fig. 7.1) when assessing the supply chain against the MSC chain-of-custody certification standard. Often this supply chain will involve a number of different organisations, transportation and storage facilities. It is up to the certifier to

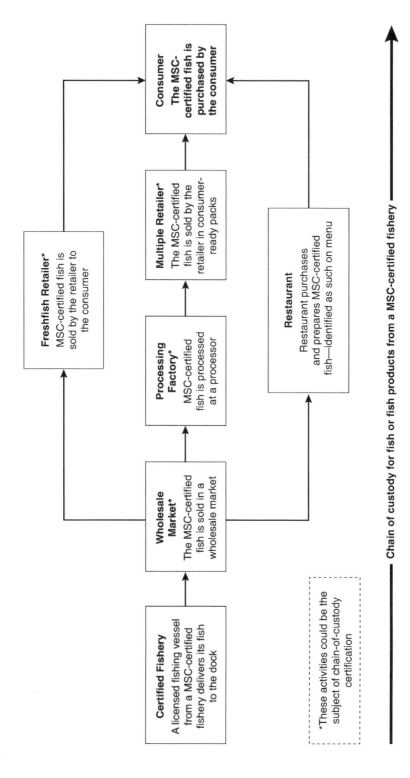

Fig. 7.1 Summary of chain of custody from an MSC-certified fishery through to the consumer.

determine which parts of the supply chain they need to visit and which can be assessed through a review of documentation. The certifier will pay particular attention to any steps in the supply chain where products from an MSC-certified fishery could be commingled with the same products from non-certified fisheries. Due to the licensing costs associated with being able to use the MSC logo on product, it is more usual for a fish processing plant to be the place at which the chain-of-custody certification assessment takes place rather than a restaurant.

7.4   The MSC chain-of-custody certification process

An organisation wishing to apply the MSC logo to fish from a certified MSC fishery must contract a MSC accredited certifier to assess the supply chain for that product. The certifier will request documentation relating to that supply chain and will usually visit one or more points in the supply chain to verify the product to which the logo will be applied originated from a MSC certified fishery.

For many years traceability within the food preparation sector has been paramount, and particularly with the European Union and US requirement for HACCP certification (see Chapter 2.4) of all food processing plants supplying those regions. Therefore, in most circumstances, verification of a supply chain is straightforward since these organisations already operate product identification and tracking systems for HACCP and other certification systems such as international quality management system standard ISO 9000 (Chapter 2.4). If an organisation already operates in compliance with the HACCP, European Food Safety Inspection Service (EFSIS), US Food and Drug Administration (FDA) or US Department of Commerce (USDC) documentation programme, it may be easier for the company to demonstrate to the certifier that it meets the MSC chain-of-custody standard. However, compliance with other programmes does not replace the requirement for MSC chain-of-custody certification, as it is possible to gain, for example, HACCP compliance, and hence permission to export, on a self-declaration basis pending the follow-up in time by the appropriate assessment authority. If the certifier can verify that the product identification and segregation systems in place are adequate to ensure that products from certified fisheries are not and cannot be commingled with products from non-certified fisheries, the organisation will normally receive a chain-of-custody certificate.

The initial certification visit usually takes about 1 to 1.5 days involving an on-site visit to the processing facility. Following the on-site audit, a report is prepared and issued to the client documenting the systems in place and compliance with MSC standards.

7.5   The MSC chain-of-custody standard-requirements

The following paragraphs summarise the compliance requirements of the MSC chain-of-custody certification standard (*www.msc.org*).

'(1)   Documented control system

- The company must have a clearly documented control system, which addresses all the section of chain-of-custody control as specified below.
- For each section the documented control system must:
  - specify the personnel responsible for control;
  - provide examples of any associated forms, records or documents;
  - specify the correct requirements for completing any associated forms, records or documents.

(2)   Confirmation of inputs

- The company must operate a system for assuring that inputs are themselves certified, if specified.
- The system must include the following requirements:
  - when the company orders MSC-endorsed fish and fish products from its suppliers, it specifies its requirement that such fish and fish products be covered by an MSC-endorsed chain-of-custody certificate;
  - when the company receives MSC-endorsed fish and fish products from its suppliers, it checks the invoices or accompanying documents to ensure that the registration code and expiry date of the chain-of-custody certificate are quoted;
  - if the company is in doubt about the validity of the chain-of-custody certificate registration code, the company checks its validity with the issuing certification body or with MSC secretariat.

(3)   Separation and/or demarcation of certified and non-certified inputs

- The company must operate a system for ensuring that when certified inputs are received they be clearly marked or otherwise identified as certified.
- Certified inputs must remain easily identifiable as certified throughout processing or manufacturing. This may be achieved by:
  - physical separation of certified and non-certified production lines;
  - temporal separation of certified and non-certified production runs.
- If certified and non-certified inputs are mixed, reliable data must be recorded which allow an independent assessor to confirm the volumes and/or weights of certified and non-certified inputs, over a specified production period.
- The outputs of processing or manufacturing of certified fish and fish products must be clearly marked or otherwise be identifiable as certified.

(4)   Secure product labelling

- The company must operate a secure system for the production and application of product labels.
- The company must accept legal responsibility for ensuring that the MSC logo pack(s) issued to the company is not used by any unauthorised users, or for unauthorised uses.
- The company must operate a system, which ensures that only its own certified fish and fish products may be labelled with the MSC name, initials or logo.

(5)   Identification of certified outputs

- Certified fish and fish products must be labelled or otherwise be identifiable in a manner that labels do not become detached during storage, handling or transport.

- The company must operate a system that allows any product sold by the company as certified to be linked to the specific sales invoice issued by the company.
- The company must operate a system to ensure that all sales invoices issued for certified fish and fish products:
  - include a description of the product(s);
  - record the volume/quantity of the product(s);
  - quote the company's correct chain-of-custody certificate registration code and expiry date.
(6) Record keeping
- The company maintains appropriate records of all inputs, processing and outputs of certified fish and fish products.
- The records are sufficient to allow an independent assessor to trace back from any given certified output to the certified inputs.
- The records are sufficient to allow an independent assessor to determine the conversion rates for the manufacture of certified outputs from given certified inputs.
- Records are maintained for a minimum of three years.'

The MSC requirements for chain-of-custody certification are relatively straightforward and documentary compliance can be illustrated by the supply chain for a fictitious Alaskan Salmon processor in the UK (Fig. 7.2). The fish is caught in Alaskan waters in July, part processed at sea, frozen into packs, shipped to Japan, containerised in Japan, and shipped to the UK. In the UK enough fish for the whole year's production arrives in November and is kept in deep frozen storage. Each day sufficient salmon is taken out of storage to satisfy that day's production schedule, and it is inspected, processed and collected by the retailer's own transport at the end of each day. The only risk of commingling in this fictitious example would be in the processing plant, which could also process farmed or wild salmon from other sources. However, in this example, the processing plant is fully occupied at this time with only Alaskan-caught salmon, consequently, the document trail is quite straightforward and could be limited to an examination of the following documentation prior to processing:

- raw material specification for Pacific salmon fillets;
- purchase order for the year's supply;
- shipping line invoice;
- US FDA health certificate for the processing ship;
- shipping line bill of lading;
- consignment insurance certificate (referencing container and customs seal number); and during and post processing:
- purchase order batch number;
- daily production schedule;
- frozen block labels;
- colour traceability labels;
- salmon frozen intake inspection and grading sheets;

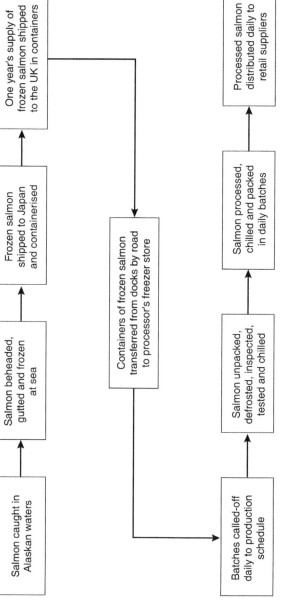

Fig. 7.2   The salmon supply chain from the Alaska fishery to a UK retail outlet.

- daily order sheet;
- daily despatch schedule;
- customer invoices;
- MSC certification codes.

7.6 How long does chain-of-custody certification last?

Once certification is achieved it is valid for five years with an on-site surveillance visit each year by the certifier. The certifier may also undertake random audits, as guaranteed by contract, at any time between surveillance visits to ensure the supply chain is operating in an appropriate manner if there is some basis for concern.

7.7 How much does chain-of-custody certification cost?

As with any certification activity, the cost of chain-of-custody certification is agreed between the certifier and the client and will depend on the size and complexity of the supply chain. For instance, if the fish product remains mostly unchanged throughout the supply chain and there are few processing sites and few stages in the supply chain where commingling can occur, then certification is likely to be straightforward. However, if the product format changes a number of times, multiple locations are involved and commingling is an issue, then the certification process will be more complicated.

A certifier will provide a specific quote prior to undertaking the work. There will also be a cost associated with the certifier's annual surveillance visits and renewal of the chain-of-custody certificate. Costs will vary by the hourly or daily rates charged by certifiers in different countries. In some cases, it may be cheaper to hire a certifier out of country and pay travel costs if the exchange rates are favourable. Once a company has received a chain-of-custody certificate, it must enter into a license agreement with the appropriate MSC entity before it is permitted to use the MSC logo.

Case Study 1:
The Western Rock Lobster
A: The Fishery and its
Assessment

8

Bruce Phillips, Trevor Ward & Chet Chaffee

8.1 Introduction

The western rock lobster (*Panulirus cygnus* George*)* occurs only in the waters of south-western Australia and supports the largest rock-lobster fishery in the world, with seasonal catches averaging 10 700 t over the period 1980–81 to 1997–8. In 2002, 594 boats in the commercial fishery were operating some 57 000 pots over the season (mid-November to the end of June), totalling an estimated 10.75 million pot-lifts each season. The 594 vessels share between AUD $200 and AUD $400 million gross per season, making it Australia's most valuable single-species fishery. Because of the high prices paid for *P. cygnus* and the good returns a fisher can expect, the stock has experienced very high and increasing exploitation since 1980. The stock is classified as being fully exploited.

During the developmental years of the fishery (1945–62) there were no restrictions on licences or gear, and the number of new entrants, boats and pots (traps) grew rapidly. By the late 1950s and early 1960s, the fishing effort was still rising rapidly but the catch had stabilised at 7–8000 t. Fishers saw their individual catches and catch rates declining and feared for their economic viability. In 1963, after representation from the fishing industry, the government introduced a policy of limited entry and closed the fishery to new entrants. In 1965 the number of pots the fishers could use was strictly controlled, limiting the total number of pots to 76 623, and the number of boats was limited to 845. Since the introduction of limited entry, both nominal fishing effort (pot-lifts) and the effectiveness of that effort continued to increase as fishers built larger vessels and worked more days per month, gaining experience and improving their gear and fish finding technology. In an attempt to offset these increases in fishing effort and efficiency, the fishery managers, in consultation with

Fig. 8.1 Western rock lobster (*Panulirus cygnus*). Photo: Department of Fisheries, Western Australia.

industry, introduced measures to shorten the fishing season, strictly define pot design and dimensions, and reduce the number of pots and boats in the fishery (Brown, 1991). These measures have been successful in slowing the rate of increase in effective fishing effort, but have not stopped it.

In addition to the commercial fishery, an active recreational fishery operates in the inshore waters (< 20 m depth) around population centres, and catches an estimated 800 t per year, equivalent to 7.7% of the catch of the commercial fishery.

8.2   The Western Australian management system

Early management measures, including limits on minimum size and the taking of egg-bearing females, were introduced in the late 1800s. Closed seasons to protect spawning stock were first introduced in 1962 and limited entry, including limits on number of boats and pots, was instituted in 1963. Use of escape gaps in pots became mandatory in 1966, with reductions in pots in 1987 and 1992, and further restrictions on retention of female lobsters in 1992. The use of output controls (quotas) has been debated, but the fishery remains input-managed.

The primary method of controlling access to western rock lobster resource in Western Australia is through licensing, empowered under the *Fish Resources Management Act* 1994. To be able to take western rock lobster on a commercial basis all the crew involved must have a commercial fishery licence (CFL). In addition, the boat must be a licenced fishery boat (LFB). Finally, a managed fishery licence (MFL) for the west coast rock lobster managed fishery must be held. Although there is no limit on the number of CFL on issue, the number of fishing boat licenses (FBL) has been falling since 1983, when the issue of new ones was frozen for all but special circumstances. The number of MFL in the western rock lobster fishery has been limited since 1963 as part of the suite of management measures for the fishery, which are expressed in a type of subsidiary legislation – the management plan – together with various regulations. These regulations are generally used to stipulate biological controls (see below) and to control the recreational fishing sector generally.

The main aim of the management system is to introduce a suite of biological, social and economic controls that maximise the benefits for the economy. Controls include maximum trap holdings and the (now defunct) '7 to 10' rule which mandated the spread of trap numbers per length of boat. The aims of these controls were to ensure that the benefits from the fishery were relatively evenly and widely distributed across the fleet. The types of input controls used in the western rock lobster fishery fall into five categories: biological, gear and season, socio-economic, compliance efficiency and other.

Biological controls are the foundation of good management in the western rock lobster fishery. They are designed to ensure that, as far as possible, immature animals are protected through the setting of legal minimum sizes. Additional protection can be given to spawning females, or even females with maximum reproductive potential, through maximum sizes, or the ban on keeping females in breeding condition. These measures do not of themselves maximise production from the fish stock but do form the underlying 'safety net' to protect the stock for future exploitation. However, it is also considered important to ensure that there is a critical biomass of adult stock so that enough eggs and young are produced each year to cope with both the vagaries of environmental fluctuations, which can effect the settlement of young lobster, and the impact of commercial and recreational fishing.

The western rock lobster fishery gear and season controls include a tightly regulated total number of traps (pots), which can be deployed by the limited number of boats, that has been periodically reduced over time. Fortunately, given the knowledge gained of the western rock lobster biological and population structure and dynamics over the last five decades, it is possible to manipulate the number of pots in use, together with associated controls, to produce a level of effort which in turn produces a relatively well predicted (and desired) total catch in any one year. This enables production to be maximised while maintaining a critical minimum adult breeding stock. Such controls have recently come under review through the National Competition Policy process. Some controls are necessary for compliance efficiency. Minimum pot holdings have been judged necessary to ensure that those participating in the fishery will, with a minimum level of competence, generate an income from the fishery and have sufficient investment in it to act as a disincentive to cheating. There are also a range of requirements to mark traps and floats, as well as the bags and crates containing the catch. In effect, the western rock lobster fishery is managed through a system of individual transferable effort, which is a surrogate for a desired catch (output from the fishery).

The principle method of consultation in the fishery is through the Rock Lobster

Industry Advisory Committee (RLIAC), a statutory ministerial advisory committee. Specifically, the functions of the RLIAC as set out in Section 29 of the *Fish Resources Management Act* 1994 are:

- to identify issues that affect the rock lobster fishing;
- to advise the minister on matters relating to the management, protection and development of the rock-lobster fisheries; and
- to advise the minister on matters relating to rock-lobster fisheries on which the views of the RLIAC are sought by the minister.

Since its establishment in 1965, RLIAC has had communication and consultation with industry as its first priority. RLIAC has been the forum in which issues (particularly those related to sustainability) have been debated and from which recommendations have flowed to the minister. In summary, RLIAC's consultative process has involved taking problems that affect the fishery and the industry generally, for example prior to 1993–4 the dangerous decline in the level of the breeding stock, and initiating, by way of discussion and management papers and meetings with industry, ways to address those problems.

8.3   Stock assessment

The success of the management measures in this fishery is demonstrated by the maintenance of high average levels of catch over an extended period of time (average of 10 700 t per season since 1980). Variations in catch from year to year are closely related to larval settlement levels three or four years earlier, and these in turn seem to be more influenced by environmental factors than by levels of spawning stock (at the current spawning stock size). In the early 1990s evidence emerged that egg production was at historically low levels (15 to 20% of unfished levels, Walters *et al.*, 1993). Even though there was no evidence of declines in recruitment at these levels, nor, in fact, any evidence of a relationship between parental-stock levels and subsequent larval settlement and recruitment to the fishery, a decision was taken to institute measures to rebuild the spawning stock to at least 25% of unfished egg production levels. Available evidence now shows the stock has recovered to about 30% of pristine egg production over the past six years (N. Hall, unpublished data). The current harvest strategy consists of a set of input controls designed to maintain the stock above 25% of pristine egg production and current effort levels would apparently stabilise the stock at about 30% of pristine egg production (Hall & Chubb, 2001). Clearly, this has been a successful strategy; harvests have continued to remain high and this effort has demonstrated to stakeholders that the fishery is responsive to perceived problems.

8.4   The MSC assessment

The first discussions about the western rock lobster fishery applying for certification under the MSC programme happened between industry representatives in Perth,

Western Australia and members of the WWF. After initial discussions, WWF invited SCS, a certifier that had notified the MSC of its intent to seek accreditation, to join in discussions to help the industry get a better understanding of the audit requirements and performance evaluations that would be required. Initially, the fishing industry in Western Australia was sceptical that the MSC programme would be beneficial. Almost a full year was required for discussions to convince the fishing industry in Western Australia that applying for the MSC programme offered more opportunities than risks.

In 1999, SCS was chosen to conduct the pre-assessment and full assessment. The client of record was the Western Australian Fishing Industry Council (WAFIC), the principal industry body in fisheries for West Australia. The (then) chief executive of WAFIC, Brett McCallum was appointed to liaise with the evaluation team to help ensure that whatever was needed in terms of information or contacts would be readily supplied. The Department of Fisheries, then known as Fisheries WA, was also on record as being supportive of the effort and available to assist in the evaluation process. This was of immeasurable help as most of the data and personnel required in the assessment were housed in the Department of Fisheries, so any lack of co-operation would have made the MSC assessment difficult to complete.

At the time of the western rock lobster assessment, the MSC had barely set up accreditation functions to identify certifiers competent to perform fisheries assessments. Since SCS had declared quite early in the process its intent to become an accredited certifier, the MSC had already reviewed SCS's qualifications and was familiar with its abilities to conduct these assessments. In addition, the MSC used the evaluation process as part of the accreditation procedures, keeping a close eye on SCS's performance throughout the project.

To develop an evaluation team, SCS contacted stakeholders from industry, the government, and conservation groups to get a list of qualified nominees who could meet the requirements set forth by the MSC. Upon gathering all the names, SCS sent the list around to the same set of stakeholders asking for comments on the viability of each nominee. From this process SCS was able to identify three scientists that were acceptable to all the stakeholders polled and that met the requirements for technical expertise and objectivity. This proved to be a critical factor in the ability of the evaluation team to conduct the assessment as it helped establish with the various stakeholder groups the transparency of the process and it led to the evaluation team containing at least one scientist that was held in high esteem by each group, thus providing access to these groups for candid interviews and discussions. The evaluation team members contracted for the project were:

- Dr. Bruce Phillips, Curtin University of Technology, Western Australia (formerly of CSIRO and a world expert on lobster biology);
- Dr. Tony Smith, CSIRO, Hobart, Tasmania (globally-recognised expert in stock assessment);
- Dr. Trevor Ward, University of Western Australia, Western Australia (formerly of CSIRO and a recognised expert in fisheries ecology and ecological impacts of fishing).

The fishery evaluation took place from 1999 to early 2000, at which time the certifier and its evaluation team determined that the fishery was in compliance with the MSC principles and criteria and recommended it for final certification. However, some factors in the fishery were determined by the evaluation team to have lower performance than expected, leading to some conditions being placed on the fishery to make improvements in the affected areas over the 5-year term of the certification. As part of the certification process, WAFIC and the Department of Fisheries signed a memorandum of understanding with the certifier agreeing to meet the conditions and fix the problems as stipulated by the certifier (and discussed below) or lose the certification. In March 2000 the western rock lobster fishery was given MSC certification. Since that time, three surveillance or monitoring efforts have been completed on the fishery to ensure that it is continuing to meet its obligation regarding the conditional requirements as well as continuing to maintain the level of management required for compliance with the MSC principles and criteria for sustainable fisheries management.

8.4.1 Conditional action requirements

Principle 2, Criteria A

Within 14 months of certification, a comprehensive and scientifically defensible assessment of the risks of the fishery and fishing operations to the ecosystem (ecological risk assessment) will be completed, based on existing knowledge, and taking into account points 2 to 5 in Criterion 2C. The assessment should consider risks of all aspects of fishing (see intent in Criterion 2B) on species (including protected and ecologically-related species), habitats, and biotic communities (see Criterion 2A). The risk assessment will identify and prioritise gaps in knowledge. The risk assessment will produce a set of prioritised risks, and strategies to address those risks, including research strategies that make maximum use of comparisons between fished and unfished areas. The risk assessment will be reviewed by independent and external expert reviewers, and be available for public comment.

Principle 2, Criteria B

Within 24 months of certification, an environmental management strategy (EMS) for the fishery will be prepared and distributed for public comment and input. The EMS will address impacts of the fishery on the environment, and will include proposed objectives, strategies, indicators and performance measures. The EMS will specify an operational plan, including implementation actions and a supporting programme of research. Future research should aim to provide information on the impacts of the fishery on the ecosystem that is at least as scientifically valid as that produced by studies of fished versus unfished areas.

Principle 2, Criteria C

Within 36 months of certification, an environmental management strategy will be incorporated effectively within the operational arrangements for the fishery.

Principle 2, Criteria D

Within 24 months of certification, there will be increased participation of the environmental community or their representatives in the decision-making processes in the fishery. This will include consultation on impending decisions, and involvement (full participation) in the decision-making processes at a range of levels in the fishery.

Principle 2, Criteria E

Within 12 months of certification, formal monitoring systems in the fishery will have improved arrangements for recording data on the by-catch of, or any other inter-actions of the fishery with, mammals, seabirds, manta rays, dolphins, or whales.

8.5   Environmental issues

The fisheries management system contained few objectives and strategies designed to assess and monitor the effects of fishing on the wider ecosystem in the fishery, or to identify any important impacts that might be occurring. At the time of the MSC assessment, there were no identified impacts which the fishery-management authority thought needed urgent attention, and evidence was provided to show that there had been a positive and reasonably effective response to issues that had arisen in the past (such as the effects of fishing in the Abrolhos Islands, a highly sensitive area of high-latitude coral reefs, and the impacts of discarded plastic straps – bands – that package the boxes of bait used in the pots). However, what was not available was a pro-grammatic strategy in the fishery to identify potential ecological risks (obvious or not) and determine if actions were needed to mitigate any problems arising through fishing.

In reviewing information during the assessment, the evaluation team did identify one sensitive issue over interactions between the fishery and Australian sea lions which breed within the fishing grounds. As Australian sea lions are at the limits of their natural distribution in the most intensively fished areas, little information was available to quantify the level of interaction of the fishery with these animals. In the MSC assessment of the fishery, the evaluation team examined information and data that suggested the effects of the fishery were not a major impact on sea lion popu-lations; however, the team also recognised the need for further assessment incor-porating additional monitoring into the conditions for continued certification as well as a priority for each annual surveillance audit.

In addition to noting the need for monitoring the sea lion interactions in the fishery, the evaluation team was concerned about the lack of a programmatic review of ecological risks in the fishery. Several conservation groups had expressed concern that little was known about the overall impacts on the ecological systems in fished areas. As a full-cost-recovery fishery, there appeared to be the chance to get caught in the circular argument that no risks are being analysed because none had been identified. As a result, there was no justification for spending money on analyses and risk reduction. To alleviate any concerns and better support the notion that ecological risks in the fishery were minimal, the evaluation team made the certification of the fishery conditional on the fishery authorities agreeing to conduct a comprehensive and scientifically defensible assessment of the risks of the fishery and fishing operations to the ecosystem (ecological risk assessment). The assessment is to consider risks of all aspects of fishing on species (including protected and ecologically-related species), habitats, and biotic communities. The risk assessment is intended to identify and prioritise gaps in knowledge and produce a set of prioritised risks, and strategies to address those risks, including research strategies that make maximum use of comparisons between fished and unfished areas. The risk assessment is to be reviewed by independent and external expert reviewers, and be available for public comment. In doing this analysis, the fishery would develop a much better set of information by which to defend the continued harvest and eliminate concerns before they became significant. The requirement for the western rock lobster fishery to complete the risk analysis is consistent with similar requirements in other well-developed fisheries, and with the increasing awareness around the world that highly exploited fisheries need to make more concerted efforts to understand their potential and ongoing risks to marine ecosystems.

8.6   Learning outcomes from the assessment process

The western rock lobster fishery was the first fishery to receive MSC certification and, as such, was recognised as a *de facto* test case for the MSC programme. Although the MSC saw this fishery as a test case, it was not officially designated as one as the industry in Western Australia paid for its own assessment and declared at the outset of the project that it was seeking certification. Since this was the first fishery project under the MSC, the scientists, managers and the industry members involved in the assessment process had no prior experience, or access to such experience, or knowledge of the requirements for achieving certification. Complicating this even further, the MSC-certification procedures were still in partial development, and certifiers were provided with only the most rudimentary guidance about how to assess a fishery to determine if it complied with the MSC standard. On top of that, the MSC had not had time to meet with the various groups in Western Australia to assure that each had an adequate knowledge of the MSC evaluation process and a full understanding of how they could interact with the evaluation team. These factors when taken together, led to a few complications in the evaluation process.

   Communicating both the process issues about the limited basis of the evolving MSC standard, and the outcomes of the fishery assessment, became highly complicated and a difficult task for the assessment team. The assessment team found that, because of the limited exposure of managers and stakeholders to both the MSC standard and the normal operational procedures for a certifier's assessment team, the consultations with stakeholders involved a mixture of briefings about what the MSC process was, what the stakeholders rôles were, views about fishery performance issues, expectations of responses from the assessment team, and continual reinforcement of the need for stakeholders and fishery managers to provide evidence to support any claims about levels of performance (either for or against certification compliance). This identified the need for the MSC to develop a more complete out-reach programme to educate governments and stakeholders in other areas where fisheries may wish to participate in the certification process.

Another key issue that unfolded during the MSC assessment of the western rock lobster fishery was that the MSC principles and criteria had to be 'operationalised' by establishing a set of performance measures (now known as indicators) and decision rules (now known as scoring guideposts). Doing this for MSC Principles 1 and 3 was generally straightforward as the performance requirements indicated by the principles was in line with best practice for fisheries in many other parts of the world. However, when operationalising MSC Principle 2, it was found that the language used by the MSC suggested a higher environmental standard in fishery performance than most other parts of the world and one that was scarcely likely to be achievable by any fishery. This meant that the evaluation team had to operationalise MSC Principle 2 with indicators and decision rules that properly reflected a level of performance that could be attributed to best practice in highly developed fisheries around the world.

The lack of preparedness of the client and the fishery managers so they could respond adequately to the nature and extent of the documentation requirements of the certifier's assessment team was another key problem during the assessment. Under the MSC programme a fishery or client would normally be expected to assemble all relevant documents and evidence in support of their case for certification prior to the arrival of the certifier's assessment team. However, this was initially a bit confusing and reflected the lack of communication between the certifier, the MSC, and the client. In the western rock lobster fishery for example, a number of aspects of the management procedures were difficult to discover in written form. Instead, the relevant management aspects of the fishery were found to be contained within a broad range of legislative instruments and voluntary agreements in the fishery, as the complete management documentation had not previously been assembled in a form suitable for audit purposes. Also, the basis for fishery performance in regard to environmental data and issues, especially issues involving marine mammals and seabirds, were particularly difficult to establish. During the project, much of the data necessary to answer environmental issues was found to reside within the jurisdiction of the Western Australian Department of Conservation and Land Management (CALM) and, therefore, was not compiled by the fishery in preparation for the assessment. Much of this was again the product of a lack of understanding by the client and fishery managers of the need to collect and provide the documents to the assessment team.

Assessment teams are not, and normally would not be, resourced sufficiently to be able to conduct their own original documentary research on a fishery. Therefore, the information required for an assessment needs to be assembled by the client or their assignees in support of the case for certification prior to the arrival of the certifier's assessment team. This project helped the MSC better understand the importance of this aspect of the programme and the need to get this clearly communicated to prospective certification applicants.

B: What Certification has Meant to the Department of Fisheries and the Industry

Peter Rogers, Ross Gould & Brett McCallum

8.7 The initial assessment process

The assessment was undertaken after considerable investigation by the rock-lobster industry, the group advising the Minister for Fisheries (the Rock Lobster Industry Advisory Committee – RLIAC), the principal commercial fishing industry body (the Western Australian Fishing Industry Council – WAFIC), the rock lobster processing sector (the Western Rock Lobster Development Association – WRLDA) and the fishery-management agency (the Department of Fisheries). The initial idea of the industry seeking MSC accreditation was also very much driven by the vision of Mr Murray France, a Western Australian rock-lobster processor.

At the end of about three years of debate considering the merits of pursuing MSC accreditation and a pilot study to gauge the likelihood of success, the Western Australian Fishing Industry Council, representing industry and government, appointed Scientific Certification Systems Inc (SCS) to undertake the certification assessment. Supporting the MSC assessment process required a significant input of time and effort by industry representatives and officers in the Department of Fisheries. In addition to supplying copies of existing reports and documents, answering questions and reviewing the assessment team's reports, the department's officers were required to write a number of new summary reports describing how the fishery operated and detailing the current status of management and research. While all the required information was fairly readily available, it took more time than anticipated to supply it in a form that was assessable and appropriate for the needs of MSC certification. Industry members also attended briefing sessions and meetings with the assessment team and hosted site visits.

It soon became apparent that the initial guidelines established by the MSC for certification were a little naively constructed and assumed that the Department of Fisheries would have collected or would have readily available comprehensive data on various ecological, social, economic and environmental issues. The assessment team recognised this data was available although not always in a form that was readily assessable or independently verifiable. Certification was subsequently approved with a series of additional requirements that the assessment team felt would strengthen the management of the fishery against the principles and criteria set by the MSC. These included requirements that an ecological risk assessment and an environmental management strategy be completed. The assessment report noted:

'Given the WRL fishery is notably deserving of certification under the MSC standards, there are still a number of factors that the certification team believes are important to address in order to strengthen the overall management of the fishery and ensure that the fishery stays on course to maintaining sustainability'.

8.8   Maintaining certification

The industry and the department have already prepared for the next stage of the process – the follow-up assessments to meet the ongoing requirements set out in the report. The initial report from the assessment team recommended that to maintain certification an ecological risk assessment and environmental management strategy should be completed. This was accepted by the MSC, industry and government. The Department of Fisheries, the Rock Lobster Industry Advisory Committee and the Western Australian Fishing Industry Council then worked together to develop a strategy to address these issues.

Although the MSC and the assessment team apparently envisaged two very separate processes, it was decided that it would be more efficient to undertake the required ecological risk assessment and environmental management strategy as one two-part project. The assessment was undertaken following the Australia-New Zealand standard for risk assessment and involved industry, the Department of Fisheries and a large number of external stakeholders. The process for consideration and sign-off of the risk assessment by the MSC-approved assessment team also required a greater level of formal justification than anticipated. The department had to develop significant levels of supporting evidence to satisfy the assessment team before some conclusions from the risk assessment (which the industry and the department had considered self-evident) were accepted.

The risk assessment generally confirmed that the rock-lobster fishery had low exposure in areas most commonly identified as fishery risks. In looking at the risk ratings in areas such as rock-lobster stock sustainability it was clear that years of data gathering, research, analysis and conservative management had ensured a soundly managed fishery where it was possible to provide accurate estimates of both stock status and annual catch. The risk assessment did raise some moderate risk issues where in some cases the Department of Fisheries could not easily demonstrate it was actively managing the issues. The identified moderate risks included the interaction between fishing and Australian sea lions and leatherback turtles, impact on corals and the disposal of domestic waste in the Abrolhos Islands. The department and industry are now actively managing all the moderate risk issues. Monitoring programmes have been bolstered across the fishery to include the interaction with 'icon' species. Research into the impacts on coral reefs has been completed and the Abrolhos Islands Management Advisory Committee is addressing the domestic waste disposal issue. Just as importantly, the department and the industry have learnt the value of formally gathering information and reporting on issues it perceives as low risk or not requiring intensive or active management.

8.9   Costs of certification

The direct costs of the assessment process were in the vicinity of AUD $200 000 with at least a further $100 000 in-kind contribution by the industry and the Department of Fisheries. The subsequent launch and promotion of the MSC certification also added a further $100 000.

The rock lobster fishery operates under a cost-recovery regime where fishermen's licence fees are used to cover the costs the Department of Fisheries incurs in managing the fishery. Therefore, much of the cost to meet the requirements for ongoing certification will be met through industry's annual licence fees. It is anticipated that the next full assessment in 2005 will cost an additional $100–$150 000. While the costs cannot be readily separated from other marketing and promotional costs, the industry itself, and particularly the processing sector, also continues to incur costs in promoting the MSC certification and labelling its product as MSC certified.

8.10   Benefits from certification

8.10.1   *Primary benefits*

Being the first major international commercial fishery to receive MSC certification has brought benefits for both the government and industry. It has enhanced the image of the rock-lobster fishery as being one of the best managed fisheries in the world and, accordingly, has helped promote the image of all Western Australian fisheries as being clean, green, well managed and at the forefront of ecological management. There is no doubt that the opportunity to be the first MSC-accredited fishery was a major incentive for the promotional plans of the fishery within Australia and throughout the world. The level of promotion of the western rock lobster and interest in the certification process has been significant.

The industry understood that, being one of the first fisheries to be certified, it would take time to realise the benefits. It also understood that when other equally high profile fisheries received MSC certification (with hopefully a great deal of media attention) awareness of the MSC process and the nature of MSC certification would improve and that, accordingly, the western rock lobster fishery would receive continual reference as being one of the first MSC certified fisheries.

Generally, rock-lobster processors see the MSC label having the most benefit in the European and USA markets where consumers are more concerned with environmental issues. An MSC label that will identify a lobster as Australian will assist but currently offers little marketing advantage in Japan. Asian consumers, generally, do not discriminate their purchases in the context of environmental issues and it is considered that an MSC label will add little value in such markets unless there is some significant promotion of the MSC principles and benefits or there is a change in consumer attitudes for other reasons.

One of the major strategies behind the MSC certification was to help the industry to re-establish itself in markets in Europe. This has proved to be successful in 2000–2002, with many countries showing renewed interest in importing significant quantities of Western Australian rock lobster. At the time the rock lobster fishery received MSC certification, it received a lot of positive media attention in North America. Unfortunately, this has probably been negated by some local USA 'copy-cat' certification systems. Some North American fisheries responded by developing their own certification systems that promoted their products as environmentally friendly. These 'copy-cat' systems lack the independent certification audits of the MSC process but probably create some confusion with consumers in the North American market and therefore may devalue the MSC certification. However, if the MSC is able to raise its profile in North America and successfully promotes the benefits of buying certified seafood, certification may also deliver advantages for Western Australian rock lobster in Canadian and US markets.

Initially, the industry has found that, in even the most 'green conscious' markets, there is very little understanding of what MSC is all about. The rock-lobster industry has found that to get any market advantage from MSC certification it has to promote both the MSC and its MSC certification. Despite the current variability in acceptance of MSC certification, many in the industry see it as providing an element of insurance by protecting them from the risk of some arbitrary ban that an import market might try to impose on an environmental-management basis. They recognise that MSC certification is likely to give them greater security in terms of maintaining hard won markets.

Another primary benefit has been the added pride of those actively involved in the Department of Fisheries and the many fishers in the industry in having achieved such worldwide recognition. The participants in the fishery are very proud of their fishery and many were extremely excited to have the opportunity to achieve a world first. It is testimony to the fact that the fishery was on the right track all along, that to achieve MSC certification there was little need to change the way that the fishery has operated for over forty years. It also reflects the spirit of trust and co-operation that has developed between the fishermen and the Department of Fisheries staff in working together over many years. While there may still sometimes be robust debate over details of how the fishery should be managed, most fishermen and department officers share a common commitment to ensuring an ecologically sustainable fishery. Perhaps as important, there exists a high level of respect for each other and an understanding that they all have an important rôle to play in responsibly managing the fishery.

8.10.2 Secondary benefits

The actual assessment process provided an opportunity for the Department of Fisheries to gather together and summarise information that gave a good overview of the fishery. This information has proved useful for a number of other purposes including providing a basis for starting the review required to meet the national government's (Environment Australia) ecological sustainable development (ESD) requirements.

The ecological risk assessment process provided numerous benefits for the department and the fishery. It provided a basis for developing and refining the risk assessments approach the department is using to meet ESD requirements across all its fisheries. By initiating engagement of all stakeholders for the first time, it also created trust and respect between the recreational fishers, commercial fishers, environmental groups and indigenous interests. The fishing industry (commercial and recreational) has learnt that the other interest groups can be supportive in helping them maintain sustainable fisheries. Similarly, the other stakeholders have learnt that generally commercial and recreational fishers are responsible and equally concerned about ensuring ecological sustainability.

The positive relationship between stakeholders built out of the risk assessment exercise has not only helped facilitate the involvement of stakeholder groups in the ESD assessment process, it has also spread over into ongoing relationships between the groups. Communication between the various groups now appears to be more frequent and positive. Following the lead from the rock-lobster fishery some commercial fishing management advisory committees (MAC) have re-evaluated their position on having wider stakeholder representation on the committee. The Demersal Gillnet and Demersal Longline Fishery MAC, for example, has moved from a position of opposing any change to the composition of its membership to recommending the appointment of an environmental or conservation representative. The rock-lobster industry itself is also discussing how it can formally involve conservationists and other stakeholders in the ongoing management of the fishery.

The involvement of wider stakeholders in the assessment process has also helped promote the image of the Department of Fisheries as a responsible and consultative manager of both the rock-lobster industry and more generally the State's fish resources. It has raised worldwide awareness of the State's fisheries and the Department of Fisheries and it has become a reference point for others wanting to follow it down the MSC-certification route. However, it also created some unrealistic expectations. For example, after the MSC certification of the western rock lobster fishery, the department received an initial wave of inquiries from overseas buyers wanting to buy other fish that simply were not available in terms of species or volumes requested.

8.11 Conclusions

Overall the experience with the MSC certification has been significant for the Western Australian commercial fishing industry and the Department of Fisheries. With the greater emphasis now placed on the requirements of ESD-based fishery management, the MSC-certification process required our management approach to be fully tested across our largest and most important fishery against these principles. Our score was high in the MSC assessment but it is accepted by all stakeholders that it can still be improved with greater emphasis placed in some areas and processes adopted to manage some risks better. These same approaches are now being applied across all the Western Australian fisheries and will, therefore, help ensure consistent and successful ESD-based fishery management.

It is acknowledged that significant market benefits are still to be realised. Although some progress has been made in Europe, the lack of general community understanding of what the MSC is and what it promotes is holding up delivery of significant benefits. However, despite some disappointment with the rate of explicit payback, it is considered unlikely that MSC will be abandoned. The industry in Western Australia believes that recognition will build as more fisheries achieve MSC certification, the MSC itself settles into a more consistent promotional strategy around the world and consumers generally become more aware of the benefits of buying MSC certified products. Moreover, the certification process and subsequent events have helped engender a spirit of commitment to the general principles of ESD and an understanding of the potential market advantages of maintaining an ecologically sustainable fishery. Therefore, even if the MSC itself were to disappear, it's unlikely that those involved in managing or working in the fishery would stop promoting the fishery and its products as 'clean and green' or stop managing the fishery with an eye to ensuring it would always be capable of maintaining MSC certification.

C: The WWF Perspective

Katherine Short

8.12 Introduction

The western rock lobster fishery was the first to be assessed by the Marine Stewardship Council (MSC). It was also the first formal introduction of the MSC to the environmental non-government organisation (NGO) community of Australia and, more particularly of Western Australia. The very essence of the MSC is its stakeholder base and WWF actively promoted engagement in the MSC assessment of the western rock lobster fishery amongst the environmental NGO in Western Australia.

Australia has a healthy and vibrant environmental NGO community and every state and territory has a conservation council, or an equivalent, termed the 'peak body' for environment and conservation organisations. In Western Australia the Conservation Council represents 68 groups including one specialist marine group, the Australian Marine Conservation Society (AMCS). Nationally, the AMCS is the only national marine conservation NGO. However, there is also an excellent federally funded programme for community awareness raising and facilitation called the Marine and Coastal Community Network that has a co-ordinator in every state and territory.

8.13 The rôle of WWF and the MSC

WWF and Unilever founded the MSC. Extensive stakeholder consultation, through a series of workshops around the world, led to the development of the MSC *Principles and Criteria for Sustainable Fishing*. The MSC is now an autonomous organisation supported by WWF and other organisations with an interest in promoting the sustainability of fisheries. WWF has three distinct rôles in relation to the MSC. These are:

(1) to support and promote the MSC mission as a means of improving fishery management;

(2) facilitate stakeholder participation in MSC fishery-certification processes;
(3) monitor the adequacy of fishery assessments and the implementation of certi-
     fication requirements by the certified body.

The first of these rôles is one that can lead to confusion among other stakeholders as it
is often incorrectly assumed that WWF and Unilever govern the MSC. In fact, neither
organisations have positions on the board, funding from either is only provided for
specific projects, and the MSC's policy development and governance direction is
vested in the board as a whole.

8.14 Supporter and promoter of the MSC

The western rock lobster fishery began the process of investigating certification in
1997, following a MSC stakeholder workshop held in Australia. The workshops, held
around the world, were the basis for the development of the MSC's *Principles and
Criteria for Sustainable Fishing*. The measures were designed to seek the widest
possible input in order to develop the standard and processes of the MSC. During this
period the western rock lobster fishery was selected as a trial fishery, accepting the
invitation to test the MSC principles and criteria. At the end of this process, the
western rock lobster fishery decided to apply to the MSC for full certification. In 1999
WWF Australia began a project to promote and establish the MSC in Oceania,
fostering awareness of it in Australia and New Zealand. This was designed to reach
stakeholders from all sectors: industry, government, research and science agencies
and environmental NGO.

8.15 Facilitator of stakeholder involvement

By 1999, certification of the western rock lobster fishery had commenced and
WWF facilitated the involvement of the main conservation organisations in Western Australia. This was achieved by WWF holding discussions in Western Australia about the MSC with the environmental NGO and the industry peak body,
the Western Australian Fishing Industry Council (WAFIC). WWF actively promoted openness, transparency and communication between the industry and NGO.
Environmental NGO expressed concerns relating to the robustness of the MSC
process, the definitions and criteria to be used, the transparency of the process, the
detail of the data to be analysed and the genuine intent of the industry to be
responsible, should they be awarded the eco-label. From WAFIC's perspective,
although the MSC required stakeholder consultation as part of the process, this
was a new and different process in terms of being inclusive and designed to seek
input from as wide a range of stakeholders as possible. Additionally, a proposal to
establish a marine park at Jurien Bay, about 150 km north of Perth, was also being
debated in Western Australia as part of the development of a network of marine
protected areas for Western Australia. Lastly, the environmental NGO in Western

Australia had a philosophical conflict with the MSC approach in that it was their view that the wording of Principle 2 prevented compliance and the fishery should be made to implement some of the requirements of certification *before* being awarded certification.

The western rock lobster fishery certification identified several important deficiencies in the understanding and management of the fishery. These were:

(1) conducting an ecological risk assessment;
(2) developing an environmental management strategy;
(3) implementing the environmental management strategy;
(4) collecting data on the by-catch of icon species; and,
(5) increased participation of the environmental community or their representatives in the decision-making processes in the fishery.

A clear lesson of the western rock lobster fishery assessment with respect to NGO involvement was that, although the certifier intended and indeed attempted to conduct a robust process, the lack of clear formalised MSC guidelines prevented a defensible process. The participants were unable to understand clearly the expectations of them and also their own expectations of the process. WWF provided a detailed critique of the stakeholder process to the MSC and a process is underway to develop formalised guidelines for certifiers to follow to enable some consistency in MSC stakeholder processes worldwide.

During the full assessment one of the groups threatened to lodge a dispute with the MSC and state they are still awaiting clarification on the status of the dispute procedure from the MSC. A dispute is the formal mechanism between the Marine Stewardship Council and a stakeholder group to resolve differences about the process and outcome of a given assessment. This also provided a lesson for the MSC, that is, the need to develop robust dispute procedures, along with the guidelines for stakeholder participation from the outset. WWF believes these should be publicly available on the MSC website (*www.msc.org*) as soon as possible.

8.16 Commentator and monitor of fishery assessments and implementation of certification requirements

WWF Australia did not provide a submission on the assessment of the western rock lobster fishery due to the assessment occurring at an early stage in the development of WWF's work on the MSC. However, subsequent to certification, WWF Australia has been actively involved in the monitoring of the implementation of the requirements of certification. The MSC provides a useful mechanism through which to achieve this through the fishery being required to conduct an ecological risk assessment and implement any of the outcomes of the risk assessment in the management of the fishery. In this context there are two key general ecological issues that WWF is particularly interested in understanding in the western rock lobster fishery:

(1)   the by-catch of Australian fur seal and sea lion pups that enter the pots seeking food and subsequently drown, and the entanglement of leatherback turtles in pot ropes;

(2)   the broader ecosystem impact of removing an average of 10 000 t of western rock lobster per year for more than 40 years.

As noted in the list of requirements on the fishery, the by-catch of icon species has a specific requirement, whereas the ecological risk assessment is the most appropriate process to ask the questions about WWF's second main point, the removal of rock-lobster biomass. Ecological risk assessment (ERA) of marine capture fisheries is a new field in fishery management and, as such, the process conducted for the western rock lobster fishery was a learning exercise for all participants and needs to be improved upon in future assessments for other MSC-certified fisheries. The ERA ranks the impacts on the icon species as 'moderate' and the risks of the removal of rock lobster to higher and lower trophic levels as 'low', however, there were no comprehensive ecological data to support this at the time and WWF requested these rankings be fully justified in the final report. Fisheries Western Australia, in part-nership with WAFIC, ran the ERA process and a final report is available.

The development and implementation of the environmental-management strategy is an ongoing process and the establishment of a new Rock Lobster Council as part of WAFIC should greatly enhance the ability of WAFIC to focus on the necessary actions. This will include an environmental representative in the decision-making processes of the fishery, an issue that has taken some time to resolve. A representative of the Conservation Council of Western Australia (CCWA) participates as an unacknowledged observer in the Rock Lobster Industry Advisory Committee, however, WWF and CCWA believe this should be extended to full participation on the Rock Lobster Industry Advisory Committee itself. This is seen as necessary to enable constructive relationships to be built and to provide a formal setting for the exchange of ideas and approaches.

The annual surveillance audits required by the MSC need to be accompanied by a formal report of the surveillance audit. This needs to be produced within a specified timeframe to keep the faith with stakeholders.

8.17   Summary overview

The MSC certification of the western rock lobster fishery is a considerable success for the fishery, resulting in new markets and new opportunities for developing rock-lobster products. WWF is still waiting to understand whether any physical marine-environmental or ecological management change has occurred as a result of certifi-cation. However, culture change was beginning to occur in both the industry and the conservation sectors. The industry has produced educational material about waste management, moved to amend the logbooks to record icon species interactions and is establishing a local MSC working group to move the implementation of MSC requirements along. Additionally, for the first time in the 50-year life of the management of this resource, it has resulted in all of the data about the fishery being collated in one place. Lastly, the basic design of the process itself and WWF's facilitation has enabled closer involvement of the local environmental community with the fishery and is thus also a 'shift'. Unfortunately, however, WWF understands there is currently a great deal of hostility between the two sectors due to the marine protected area debate and changing personnel.

Following certification, WWF's active facilitation was reduced and the engagement of the conservation sector with the industry declined as a consequence with some of the momentum created now lost. WWF believes this is cause for concern because of the lost opportunity to capitalise on the initial relationship building that occurred. Another lesson of the western rock lobster certification is that third-party local facilitation and formalised structures for this new way of working are necessary at the assessment point and then throughout the life of the certifications.

From the perspective of industry culture change and marketing benefits, it has become apparent that a detailed internal industry strategy is also necessary post-certification to promote the benefits of certification to the industry, to maintain momentum and to entrench the achievement of the successful certification. This is critical throughout the whole extent of the fishery to maintain industry interest, secure funds for initiatives and to produce real lasting change at all levels.

D: An Unsatisfactory Encounter with the MSC – a Conservation Perspective

David Sutton

After all the fanfare, and despite positive progress in marine conservation attributable to the efforts of the Marine Stewardship Council (MSC), the conclusion has to be the same: the MSC was in error in certifying Western Australia's western rock lobster fishery. It certified the fishery as meeting a sustainability standard which it has never been found to achieve, and its certification is both inaccurate and misleading. The consequences for the MSC and its supporters are, firstly, that they risk their credibility for honesty in labelling and hence, risk undermining their very objectives. Secondly, and equally damaging, is the potential for certification to undermine other conservation initiatives, and thirdly, and possibly most disturbing, is the MSC's apparent determination to ignore the problem.

It should be said from the outset that what follows is not about the MSC assessment of management of the western rock lobster fishery. This must be one of the more secure fisheries in the world. Fortuitously the species is particularly fecund and valuable, the responsible management agency has sufficiently good stock indicators to allow relatively accurate predictions to be made about harvestable stock levels into the near future, and mechanisms exist to control catch where necessary. In this sense, the western rock lobster fishery is well-managed and deserves appropriate recognition. However, knowledge of its environmental impacts and hence its ecological sustainability is far from certain and no certification should imply otherwise if it is to be credible. Herein lies the problem for the MSC.

Conservation groups in Western Australia became aware of the MSC in mid 1999 through the announcement that the western rock lobster fishery was the first to be assessed by this new body. Some of us were aware of the Forest Stewardship Council and its objective of promoting ecologically-sound forestry practices, so we came to the initial meeting with the western rock lobster certifier with interest and open but critical minds. It is worthwhile focussing briefly on the 'critical' aspect of our approach.

Western Australia has approximately 13 000 km of coastline and a vast extent of waters extending out to the limit of the Australian exclusive economic zone. It has one of the world's most singular marine environments and an incredibly high proportion of endemic but, as yet, poorly studied and described marine biota. Thus, it is not unusual that 50% of some phyla (e.g. sponges) found in habitats where the western rock lobster occurs are new to science. Only a small and unrepresentative part of this environment is encompassed in marine conservation reserve areas, and an inconsequential part has fully protected sanctuary status. The conservation sector has argued for many years that this urgently requires rectifying, in no small part because Western Australia waters have been subject to intensive recreational and commercial fishing pressure over many decades, including by the very lucrative western rock lobster fishery. Both these sectors clearly affect the marine environment but the full extent of their effects are yet to be discovered and the very real possibility exists that, individually and cumulatively, these sectors have been responsible for major changes to the marine ecosystem of Western Australia. A comprehensive, adequate and representative system of sanctuary areas is an essential core component in managing for sustainability of marine ecosystems. Amongst other things, sanctuary areas have important rôles in protection of biodiversity and ecological structure and function; they provide the potential for restoration of representative marine habitats and biodiversity, and in so doing provide a window on pre-exploitation condition; they allow for ongoing evolution without the undue influence of human activities, and importantly; provide scientific reference areas enabling change and impact to be monitored.

Both the commercial and recreational fishing sectors have a history of opposition to establishment of fully protected marine conservation areas in Western Australia, and marine conservation areas generally. At the same time, the MSC process was being initiated for the western rock lobster fishery, the conservation sector was watching commercial fishery representatives protecting their existing access to almost the entire central and south coast marine environment of Western Australia by undermining the establishment of all but trivial sanctuary areas in a multiple-use marine reserve proposed for Jurien Bay, about 150 km north of Perth. Jurien Bay is near the centre of the western rock lobster fishery, which spans about 1800 km of the Western Australia coastline. Given 40–50 years of western rock lobster fishery impact on the marine environment, the absence of any meaningful ecological baseline data on this impact currently or in the past, the absence of non-fished areas to act as reference sites and the western rock lobster fishery opposition to their establishment, it is hardly surprising that the conservation sector would take a careful and critical look at any process purporting to assess the fishery's sustainability.

Despite these reservations, we saw in the MSC possibilities for influencing the fishing industry generally towards ecological sustainability. We also saw the potential status, credibility and enormous economic advantage that certification by the MSC could deliver to the industry, and realising we were the conservation sector test pilots for this new approach to marine conservation, took our rôle seriously in getting it right. We continue to do so. From the first meeting the alarm bells rang and have not stopped ringing, and our concerns are not restricted to the western rock lobster but to all fisheries assessed by the MSC as it stands at present. At the core of the issue is what MSC certification means.

So what can a fishery claim about its environmental effects once it has been certified? The answer to this lies in the MSC *Principles and Criteria for Sustainable Fishing* wherein 'Independent bodies will certify whether a fishery *complies to MSC's sustainability standard*', and '...*fisheries which conform to the Principles and Criteria* will be eligible for certification...' and 'Products from fisheries *meeting the standard* are eligible to use the MSC logo...' (MSC glossy public brochure titled Our empty seas: *A global problem – a global solution*), and 'Consumers, concerned about over-fishing and its environmental and social consequences will increasingly be able to choose seafood products which have been *independently assessed against the MSC Standard* and labelled to prove it' (*www.msc.org*). It is statements such as these which can leave no doubt in anyone's mind that for a fishery to be certified it must meet the MSC sustainability standard, i.e. the principles and criteria.

Our assertion to the MSC chief executive, the fisheries director, the certifier and the evaluation team for the western rock lobster fishery remained consistent and is unchanged: no fishery is able to legitimately claim, or to have claimed on its behalf, that it meets the critical environmental impact criteria (1 and 2) for Principle 2. It is worthwhile to reflect on what these require a fishery to demonstrate about its activities and to reflect on the knowledge base necessary to enable assessment of the veracity of any claim in this regard. The two criteria we draw attention to are reproduced below, and require that:

(1)   the fishery is conducted in a way that maintains natural functional relationships among species and should not lead to trophic cascades or ecosystem state changes;
(2)   the fishery is conducted in a manner that does not threaten biological diversity at the genetic, species or population levels and avoids or minimises mortality of, or injuries to endangered, threatened or protected species.

The second of these criteria requires that the biological diversity (genetic, species and population) is known throughout the habitats where the lobster occur. The first requires detailed knowledge of the ecological role of the target species and its relative importance in the ecosystem, the natural functional relationships amongst species (including but not restricted to the target species) in these habitats as well as an understanding of factors determining how trophic cascades or ecosystem state changes might come about in the particular ecosystems involved. Any fishery able to demonstrate conclusively that it meets both these criteria would be quite justifiably hailed by the conservation sector. However, in the case of both criteria, the current knowledge available for attempting to assess any fishery against these criteria is superficial, rudimentary, speculative or non-existent, and quite inade-quate for the task. The evaluation team for the western rock lobster assessment concurred with this view. It is exasperating and an ironic indictment of the process that research into many of the above matters, required to be undertaken by the fishery if it is to maintain its MSC certification as ecologically sustainable is precisely what is necessary to help decide in the first place whether it is or is not ecologically sustainable.

The reason the evaluation team could agree with our appraisal yet were able to continue with their assessment of the fishery became apparent through correspondence with the MSC. The MSC Fisheries Director informed us that in fact the MSC did 'not require that a fishery score 100% [against the standard] in order to be certified ... Rather, we require that a fishery achieve a sufficiently high score against the standard' and 'In order to be certified ... the fishery must achieve at least an 80% score against each Principle, although it does not necessarily need to achieve an 80% score against each criterion within a Principle' (correspondence dated 24/12/99). Subsequently, the MSC Accreditation Officer, writing on behalf of the Fishery Director, informed us that 'Principle 2 was never meant to be an absolute standard' (correspondence dated 17/4/00). Not very comforting revelations in view of publicly available statements to the contrary and the chief executive's assurance to us that certification required that a fishery meet all principles and criteria. Compounding this, it became apparent that the fishery was not really being assessed by the evaluation team against the criteria stated in the MSC standard, but against 'performance criteria' developed by the evaluation team. I will have more to say about these performance criteria later, but suffice it to say they are very comprehensive in assessing the fishery management system. However, as we pointed out to the MSC and to the evaluation team, the performance criteria in no way equate in substance to the criteria of Principle 2 and it would be absurd to suggest that a fishery favourably assessed according to the performance criteria could, therefore, be said to satisfy the ecological criteria of Principle 2 as stated above. Anyone who cares to confirm this should do their own comparison of the performance criteria against the criteria defining the MSC sustainability standard.

The assertion that the MSC principles and criteria and the performance criteria for the western rock lobster fishery do not measure the same thing (i.e. ecological sustainability) is clearly reinforced in the *Public Summary for the MSC Certification of the Western Rock Lobster Fishery Western Australia*. In this, the certifier states that, in relation to ecological concerns, the fishery receives a 'pass' for nothing more than 'minimising ecosystem impacts'. Minimising ecosystem impacts could hardly be said to be the same as ecological sustainability, as defined by the MSC standard.

Is this just semantics? Should we care whether the assessed fisheries meet the MSC standard or not, as long as they are being moved towards ecological sustainability? Clearly we believe we should care, for both ethical and practical reasons. The issue of principle (that word again!) is a good starting point and sufficient in itself: few people would agree that the ends justify the means. However, there are good practical reasons as well. Honesty in labelling is one: no product should have an unwarranted status in the market place. Credibility is another: an industry exploiting natural resources should only get credit for environmental achievements which can be confirmed. A further and critical practical reason is the deliberate or unthinking use to which certification might be put in undermining other conservation strategies, such as marine conservation areas. Why should industry, fishery managers or politicians support the establishment of areas closed to fishing activities when they can say the activity concerned is certified sustainable, having no impact on biodiversity or on marine structure, function or integrity? What happens as more fisheries operating in the same ecosystem are similarly certified despite enormous gaps remaining in knowledge of their individual and cumulative impacts let alone the ecosystem generally? As noted above, the commercial sector, supported by fishery managers, has resisted marine protected areas despite the fact that they are the only adequate scientific control sites that have the potential to determine fishery impacts. How a fishery could be considered sustainable and certified as such in the complete absence of such a basic and irreplaceable mechanism essential to assessing impact defies belief, credibility and acceptability. To its credit, the terrestrial equivalent of the MSC, the Forest Stewardship Council, at least has a criterion requiring that the presence of fully protected areas be a core element in any sustainability assessment.

In this regard it is interesting to note that following certification of the western rock lobster fishery by the MSC, the Director of Fisheries Research for the Department of Fisheries in Western Australia unambiguously advised a national meeting of Australian fisheries and environment managers, bureaucrats, marine scientists, industry and NGO representatives that the MSC has certified that the western rock lobster fishery is sustainable (*ESD and Fisheries: What, Why, How and When – a Stakeholders' Workshop* – 23–24 March 2000, Geelong, Victoria). Further, in his column *From the Minister* (*Fisheries in Western Australia*, March 2000, Government of Western Australia), the Minister for Primary Industries and Fisheries, Monty House, stated that the western rock lobster fishery is 'the first in the world to be awarded international MSC certification as a sustainable, well managed fishery'. It may be well managed but no one is in a position to say it is [ecologically] sustainable as defined by the MSC standard.

How has this problem for the MSC come about? In no forum or correspondence have our concerns been allayed or our claims of inaccurate and misleading assessment against the criteria for Principle 2 been refuted or flawed. In fact, the contrary has occurred. In response to our correspondence suggesting ways for the MSC to resolve this situation (see below), the MSC Fisheries Director (writing on behalf of the chief executive) noted that while he felt major impacts of the western rock lobster fishery could not be identified: 'We also agree that the state of knowledge about fisheries being considered for certification – and particularly their impacts on marine ecosystems and biodiversity – is often less than ideal'. Subsequently the accreditation officer (responding on behalf of the fisheries director) noted that at the MSC Senior Advisers Group meeting and at the MSC Certifiers Workshop (both held March 2000): 'There was acceptance though that there may be some ambiguity in the wording [of Principle 2 and its criteria] and it was agreed that there was a need to modify the present wording to remove any such ambiguity at the next MSC Standards Council Meeting'. We were subsequently informed by the certifier that he believed '... the MSC Standards Committee after considering this issue carefully ruled that the wording was to stay the same' and that '... the MSC Board of Directors supported this ruling'. As of 2002, the MSC website shows the present wording to have not changed. No explanation or justification was ever forthcoming to us despite our involvement and evident concerns, and I think we must assume that the MSC sees the problem but has no way of correcting the flaw without (in its view) losing professional credibility, and is loathe to do so. Thus, it is hard to avoid the conclusion that the MSC chooses to place greater importance on its credibility with those industries and agencies exploiting marine natural resources above its own principles and criteria, and its credibility in the eyes of the broader community. As it certifies more fisheries on the basis of the present Principle 2, it is compounding the error, continuing to mislead the public and ultimately making it harder to repair the damage.

What can and should be done? The simple resolution is to admit the error and change the wording of Criteria 1 and 2 (Principle 2) – simple, effective, clean and honest. Here is the rewording we proposed to the MSC:

(1) The fishery is conducted in a way that aims to (i) maintain natural functional relationships among species and (ii) minimise the possibility of causing trophic cascades or ecosystem state changes.
(2) The fishery is conducted in a manner that aims to (i) maintain biological diversity at the genetic, species and population levels and (ii) avoid or minimises mortality of, or injuries to non-target, endangered, threatened or protected species.

A small change, but enormously significant changes in emphasis. These are criteria which allow for assessment of the management of the fishery and realistically reflect what the evaluation team's performance criteria are actually capable of measuring. Using the above criteria to define the standard, and performance criteria which are their direct measure, it is possible to assess, judge and certify the evolving management of a fishery without the need for, or inaccurate claim to, certainty of ecological outcome. Without the need for misleading claims, certification against these criteria can provide confidence that the management of a fishery is moving it towards understanding and minimising its ecological impacts and progressing towards ecological sustainability.

The credibility and benefit gained by the fishery for environmental achievement would be in no way diminished, and certification would still act as a goal and incentive for other or competing fisheries. This approach would allow for and encourage ever-improving environmental standards based on definable and assessable management standards. It would eliminate ambiguity and inconsistencies, maintain faith with community expectations of accountability and also minimise potential for future criticism of the process. Finally it is worthwhile pointing to one concern we hold in common with the MSC, namely that 'Some claims [about sustainability] are deceptive and confuse consumers . . .' (MSC brochure *Our Empty Seas*, page headed 'Consumer awareness/Business sense'). We believe the MSC needs to change its own processes, deliberations, decisions and claims to be above such criticism.

Case Study 2:
The Alaska Salmon
A: The Commercial Fisheries

9

Chet Chaffee

9.1 Historic production of the salmon fishery

The Alaska salmon fisheries occur within the US territorial waters adjacent to the coast of the State of Alaska. They target five species of Pacific salmon: sockeye (*Oncorhynchus nerka*); chum (*O. keta*); chinook (*O. tshawytscha*); coho (*O. kisutch*); and pink (*O. gorbuscha*). Salmon are harvested by nets (drift and set gillnets, purse seine) and by trolling. The fisheries occur within management districts delineated by the Board of Fisheries (BOF) and are managed by the biological staff of the Alaska Department of Fish and Game (ADF&G). Salmon has been caught and processed in Alaska since the 1800s. After purchasing Alaska from Russia in 1867, the United States made the area a customs district under the Treasury Department and the first salmon saltery was opened in 1868. The first salmon cannery was opened in 1878 and by 1920 there were 160 salmon canneries in operation in Alaska.

Salmon fisheries productivity and fisheries management in Alaska began to significantly change in 1959 when Alaska became a state. At statehood the average annual harvest had fallen to about 25 million salmon, the lowest since 1900. Harvests generally improved after 1959 to about 40 million salmon per year, a trend that lasted for a period of about 12 years. However, in 1967 and again from 1972–5 poor harvests occurred. The poor productivity of salmon was thought to be attributable to lower survival during severe winter weather. Catches since then have been increasingly robust. This improvement has been largely attributed to a long-term improvement in the biological productivity of the ocean waters in the Gulf of Alaska. However, other management factors at play over the years have probably also played a role in the continuing success. Significant contributions include the elimination of high-seas drift-net fishing, the implementation of the *Magnusen Fisheries Management and Conservation Act* 1976, and improved fisheries management on the part of ADF&G.

Catches in the past decade have been generally above 190 million salmon with ex-vessel values exceeding US$400 million annually. Numerically, pink salmon predominate in the harvest, comprising more than one half of the statewide harvest. Roughly one fourth of the harvest is sockeye salmon, followed by chum, coho and chinook salmon. In product value, sockeye salmon have always been the primary species. At the time of statehood, essentially the entire harvest was canned or salted. In recent years more than 80% has been sold whole or eviscerated. Seventy-five per cent of the fresh or frozen product is exported with Japan purchasing about 80 percent of that. The canned product is sold primarily within Europe and the United States.

9.2   Alaska salmon management

At statehood (1959), Alaska assumed control of its fisheries and mandated their sustainable management in the Alaska constitution. In Article VIII of the state constitution there is an entire segment dedicated to management of natural resources, with several sections specifically pertinent to the management and conservation of salmon.

Alaska Department of Fish and Game was created by the state legislature and charged to: 'manage, protect, maintain, improve, and extend the fish, game and aquatic plant resources of the state . . .'. The legislature also established the Alaska Board of Fisheries, a seven-member panel appointed by the governor and confirmed by the state legislature with authority to establish open and closed seasons, set harvest limits, establish the methods and means employed in fishing, manage and improve watersheds and habitats, and regulate fishing. The management system in place in Alaska is designed to allow for daily management decisions by biologists qualified under the state programme as area management biologists. Each area management biologist has additional professional and technical support personnel to work on the management of fisheries. Under state regulations, an area management biologist can issue in-season management changes to regulations that can be placed into effect immediately and carry the full force and effect of law. These decisions are made through careful consideration of fishery-dependent and independent data that are collected throughout the season.

Two significant mechanisms are used to control harvest rates on Alaska salmon; limited entry and escapement goals. The state instituted limited entry in 1973. Within each management region legal gear-types were defined and permits issued to individuals who could demonstrate a history of participation within that segment of the fishery. Escapement goals set the requirements for the number of salmon that must pass through to spawning sites to maintain a sustained harvest as defined by ADF&G, thus dictating annually the level of harvest available after biologists define the strength of the salmon runs. Another aspect of fishery management in Alaska has been the implementation of salmon hatcheries to enhance runs in specific regions in the state. State hatcheries were constructed starting in the late 1970s. However, laws were later passed to allow non-profit corporations to develop private hatcheries. By the mid 1980s even the public facilities were turned over to private non-profit corporations. Now, salmon provided by hatcheries in some areas exceeds wild fish runs significantly.

Salmon fishery management in Alaska is now complicated by management input from both state and federal agencies. In 1999, the US government declared jurisdiction over the management of subsistence fisheries in waters claimed by the federal government. The effect being that the federal government has the authority to assure adequate up-river returns to support subsistence fishing. While both state and federal authorities are working together, tension between the two will continue to exist as each authority has a different set of objectives. The federal government is specifically interested in maintaining subsistence fisheries and has no mandate to maintain commercial harvests, whereas, ADF&G has the mandate to maintain both commercial and subsistence harvests as well as the long-term sustainability of salmon runs.

9.3   The MSC assessment

As discussed in earlier chapters, Unilever and WWF initiated the MSC process. From the very beginning, both organisations wanted to create a programme that took into consideration the views of fishery stakeholders from around the globe. To accomplish this task, one of the first actions taken by the founders was to develop a draft standard and then conduct workshops at various geographic locations around the globe to seek comment on the standard and proposed methods for evaluating fisheries. Among the numerous suggestions at the regional workshops was the need to test the robustness of the standard and the proposed certification methodologies using test cases. Accordingly, the MSC sought volunteer fisheries in an effort to identify 2–3 fishery projects that would be appropriate for the test cases.

Representatives from Alaska first put forth the idea that the MSC should consider Alaska salmon for a test case. The proposal was based on ADF&G having in-season management, healthy runs over decades, numerous large-scale fisheries under a single management authority, and some of the best-managed fisheries anywhere in the world. As a test case, it was suggested that the MSC would be able to determine if its system was robust enough to handle large-scale fisheries projects and multi-species fisheries, as well as being adaptable to fisheries that were not strictly marine-based.

In 1998, the MSC began discussions with a working group formed by ADF&G comprising fishers, processors, government managers, and conservation groups. The working group was formed to discuss Alaska's participation in the MSC initiative should the MSC decide that Alaska salmon fisheries would be an acceptable test case. Although the initial suggestion came from Alaska, a number of people and organisations in Alaska voiced concerns over the idea of using commercial Alaska salmon fisheries as a test case. After more than a year, the discussions of concerns boiled down to one problem – allocation. Processors and fishers alike did not want to see one gear type, or one fishery gain an advantage over others through the MSC process. ADF&Gs working group decided that they would generally support Alaska salmon fisheries as a test case project under specific circumstances:

- the project would include all species, all gear types;
- the project report, when written, would not identify individual fisheries problems, but would talk about species groupings to avoid issues that may affect arguments about allocation;
- the MSC would pay for the test case;
- the test case could be converted to an official certification evaluation if requested by ADF&G.

In other words, Alaska would volunteer staff time and expertise to answer questions from an accredited MSC certifier and to compile the necessary information for the evaluation. In return, it would get a report on whether Alaska salmon fisheries successfully comply with the MSC standard and the report would either say the fisheries passed or failed as an aggregate whole, rather than break out problems of individual fisheries that could be used later in battles over allocation.

As stated above, the evaluation of Alaska's commercial salmon fisheries included more than one species, and among each species a large number of almost independent populations. Each of the five species of salmon in Alaska is not a single interbreeding population, but rather a metapopulation of individual spawning stocks, each stock returning to spawn in the same river each year. The straying rate is typically very low (< 5–10%), making these populations functionally independent. Because of the homing ability of salmon and the ubiquitous number of spawning locations in the rivers and streams of the state of Alaska, tens of thousands of individual spawning populations prevail. Since most methods in fishery management are based on management of a single, inter-breeding population, managing tens of thousands of spawning stocks is a daunting task. As a consequence of the situation, a decision was made to manage salmon as geographic or regional groups, typically of species and populations on which a commercial fishery has historically been targeted. While the strategy employed to manage these aggregates is uniform across the state, the differing characteristics of the runs, their size and value, results in differing effort in data gathering and analyses of those data throughout the State of Alaska.

The accredited certifier chosen for the project was Scientific Certification Systems Inc. (SCS) of Oakland, California. Recognising that the large number of populations from each of the five species and the variety of harvesting modes made Alaska salmon fisheries difficult to assess for sustainability and compliance with the MSC principles, SCS chose carefully to select individual experts to form the assessment team. Through several months of consultations with ADF&G, the working group, and stakeholders, SCS selected three experts – Dr. Lee Alverson of Natural Resource Consultants (Seattle, Washington, USA), Dr. Louis Botsford (University of California, Davis, California, USA), and Mr. Paul Krasnowski (retired ADF&G biologist). Each expert brought significant knowledge of salmon biology and management to the test-case project such that the learning curve, compilation, and review of information would be made easier. Dr. Chet Chaffee of SCS managed the assessment team's activities to ensure compliance with MSC requirements. While Dr. Chaffee participated in all substantive discussions, meetings, and interviews, he did not provide input to the final scores assigned in the fishery evaluation. Instead, Dr. Chaffee facilitated the team's scoring discussions to make sure the three experts chosen had properly and objectively come to consensus on the scores assigned to the various fisheries.

Early in the process, the assessment team recognised that salmon are inherently easier to manage on a sustainable basis than many other species. Individual populations spawn only in a specific area each year, and do so in a way that makes monitoring of the spawning population much easier than estimating spawning abundance in other fisheries. However, the team also recognised the problems with the test-case mandate of looking at all species in all areas. The resulting determination by the assessment team of how to conduct the assessment is best illustrated from two passages in the final report:

'. . . it is well known that when several spawning populations are managed as a unit, it is possible for less productive populations to be lost (e.g. Hilborn and Walters, 1992). If abundances of all populations in the group are not monitored, managing a group of populations increases uncertainty regarding sustainability of individual populations. In Alaska, both the amount of data that is gathered on each individual population and analyses of these data to determine management vary from population to population.'

and

'The heterogeneity of management implementation over tens of thousands of populations from five species presented difficulties in assessing this "fishery" according to MSC principles because MSC principles and criteria for sustainability are largely based on management at the (single) population level. This difficulty is essentially a mismatch of scale, or more accurately, of level of organization. We could have required that Alaska demonstrate that all individual-fished salmon populations in Alaska were individually monitored and well managed. We did not take this approach because it would essentially be holding Alaska to a higher standard than others. Many other fisheries (e.g., the Australian western rock lobster fishery) target populations that are also really metapopulations of individual benthic subpopulations linked by a dispersing larval phase. The difference is that the coupling between subpopulations is substantially greater, resulting in an almost completely mixed, interbreeding population.'

Instead we decided to tailor our assessment of sustainability to the level of aggregation used by the State of Alaska but to maintain a concern for the populations within each aggregation. Where individual populations are managed directly, we applied the MSC population criteria directly. Where populations are managed in groups, we asked how individual populations are accounted for. We do not require that abundance of each be monitored and used in management, but we do require that the number and average abundance of each population be roughly known and that information be available on the number and abundance of populations that are assessed and managed in each different way. The difference between this and other assessments is that: whereas in most fisheries, the assessment of sustainability consists of judging *how close* all of the data gathering, estimation and stock assessment come to making *the* population sustainable, assessment of Alaska salmon for sustainability involves judging *how many* of the individual populations on which data-gathering, estimation and stock-assessment efforts are focused in the aggregate, are managed safely enough to make this 'fishery' on five species and tens of thousands of populations sustainable.

One last major issue in the assessment was the fact that some of the salmon caught in Alaskan waters originate in other jurisdictions, hence, they require special consideration in assessing sustainability. Several species, particularly chinook and coho taken in the troll fishery in south-east Alaska (SEAK) originate in Canadian rivers. Chinook salmon also taken in the SEAK troll fishery that originate in the contiguous US and are listed under the US *Endangered Species Act*. Also, several species taken in fisheries on the Yukon River spawn in Canadian rivers. Again, the assessment team's acknowledgement of this issue and its importance is best illustrated by a passage from the final assessment report:

'While these stocks originate outside Alaskan waters, and primary management responsibility lies in the hands of others, the assessment team's interpretation of the MSC guidelines is that for the State of Alaska to have their salmon fisheries certified as sustainable, it is their responsibility to fish only in sustainable fisheries regardless of the origin of the stock. We thus apply the same standards to these fisheries as the fisheries on Alaskan populations. The difference is that since these fisheries are managed by agreements (i.e. treaties, consultations under the US ESA), we focus on whether the terms of those agreements lead to sustainable fishing, and whether Alaska is living up to the agreements.'

9.4   Environmental issues

Several specific environmental issues were raised in the assessment. First and foremost was whether salmon habitat was being substantially degraded and what control ADF&G had for mitigating these impacts. The assessment team made specific efforts to gather information on ADF&G's rôle in mitigating negative effects to salmon habitat by forestry, urbanisation, mining, agriculture, and other extrinsic factors not in the direct management control of ADF&G.

The assessment team was also concerned about potential shifts in ocean productivity such that salmon survival would be reduced to levels prior to the 1970s when oceanic regime shifts improved conditions for salmon. The cause for concern was based on the increasing awareness that such large-scale changes occur repetitively on decadal timescales, and that a reversal of this increase in conditions beneficial to Alaska salmon could occur at any time. As the assessment team noted: 'The most significant question regarding sustainability of Alaska salmon fisheries, therefore, is not how much has catch increased over the past twenty years, but rather how well would the management system respond to a downturn in ocean conditions?'

The potential effects of hatcheries on salmon survival and salmon habitat were also identified by the assessment team and conservation groups. The concerns raised fell into two broad categories: the effect of returning hatchery fish mixing with natural stocks complicating harvest management and escapement, and competition for resources by hatchery fish. Hatchery stocks are able to withstand very high exploitation rates that may exceed those tolerable by a stock that spawns under natural conditions. Therefore, harvest of mixed hatchery and wild stocks can be a difficult task to manage with the idea of maximising harvest while still protecting naturally spawning stocks. The release of large numbers of hatchery juveniles into near shore rearing areas may have an effect on the growth and survival of natural stocks through competition for food. In addition, straying and spawning of hatchery-origin fish into natural spawning areas may affect fitness and productivity of wild populations.

9.5   Stakeholder concerns

Comments from stakeholders in Alaska were minimal, and mostly on the positive side. Industry stakeholders were quite supportive of ADF&G's management of salmon. Industry's main concerns seemed to centre around how an independent evaluation team would consider mixed-stock fisheries and hatchery-augmented harvests. Government agencies outside the jurisdiction of ADF&G were quick to point out that they believed ADF&G is doing a good job at interfacing with all the necessary agencies to protect and conserve salmon populations and habitat.

The conservation sector was surprisingly quiet during the assessment process. Although the certifier (SCS) contacted the local, regional, and national groups known to be concerned with salmon management, few had anything to say, and what was said was generally favourable. The few concerns voiced at the time of the assessment were by the Audubon Society, the Sierra Club of British Columbia, and the Canadian fishing industry. The concerns raised were:

- the influence of salmon hatcheries on the genetic integrity of wild salmon stocks;
- the ecological effect from adding thousands of additional salmon fry into specific areas and their effect through competition for the plankton food source;
- the adequacy of marking of hatchery salmon to understand their influence in Prince William Sound and other areas;
- interception of Pacific salmon destined for Canada and other coastal areas;
- concern that the state budget for salmon management has been cut to the point that there is little to no room for research and development into new management strategies and techniques.

All of these concerns became significant issues for research and review by the assessment team.

9.6   Outcomes from the assessment process

Although the Alaska salmon test-case assessment project was the third evaluation to be conducted under the MSC programme, it was essentially conducted concurrent with the first two projects on Thames herring and Western Australia rock lobster. This meant that the assessment team suffered many of the same obstacles as the other two assessment teams:

- there was little background or training provided by the MSC;
- the certification procedures were still in a developmental phase with little explanatory material available to provide to ADF&G and the various stakeholders in the process;
- the assessment team had to start from scratch in developing the performance indicators and scoring guides;
- due to a lack of exposure to the MSC process, the fishery managers and scientists were not well prepared when asked to respond to extensive requests for documentation proving compliance with the MSC standards.

In addition, and as noted above, the project took on the enormous task of examining a large number of mixed-stock fisheries and was constrained by agreements for conducting a test case.

It became clear after the completion of the test case that the assessment team's findings were that, overall, the commercial salmon fisheries in Alaska represented fisheries that were in general compliance with the MSC principles and criteria. As a result, ADF&G opted to have the test case extended into an official certification evaluation. This entailed increased stakeholder consultations in Canada, as well as further investigation and review of both hatchery-related issues and management of trans-boundary stocks, as investigation into both of these issues was only partially completed in the test-case project. At the end of the certification evaluation, the salmon fisheries in Alaska, as a whole, were recommended for certification and certified by SCS in October 2000. The certification was granted only after ADF&G agreed in writing to make specific improvements in the fishery (referred to as conditions or requirements for continued certification in the final evaluation report).

No official disputes were filed against the findings of the certifier, due at least in large part to the fact that groups with concerns were unaware of the ability to dispute the findings. Subsequent to the certification of Alaska salmon, a number of groups from the conservation sector and Canada have expressed some concerns about specific fisheries in Alaska. The assessment team is looking into these concerns as part of its annual surveillance and monitoring programme – a requirement under the MSC initiative. Moreover, the uneasiness of conservation groups with the findings in this project and others has led to the MSC revising and strengthening its dispute resolution procedures (see Chapter 6).

Post-test case and certification, the certifier (SCS) provided the MSC with its recommendations for future projects. One recommendation was to avoid projects so large that they are inherently too large to conduct as single projects. These projects limit the extent to which an assessment team can look into specific issues in detail. Another recommendation was to develop a base-set of performance indicators that could be used as the same starting point for all certifiers in all fisheries. This would, by its very nature, help promote consistent results between assessments. Lastly, it became clear that the stakeholder consultation process needed to be improved, as this was an important aspect of the MSC evaluation process since it affords the evaluation teams the only opportunity to hear about concerns regarding fishery management. The rest of the evaluation process is focused on reviewing information submitted by the client or fishery as proof of good fishery management and, therefore, does not provide the counter-arguments made available by concerned stakeholders.

B: A Fishery Perspective

Robert Bosworth

9.7 Introduction

Alaskans are fiercely proud of their salmon. Alaskans catch and eat a lot of salmon, and during the summer salmon runs they are a big part of what people talk about. As a wild and renewable resource, salmon make a huge contribution to Alaska's economy. They are a force of nature. They keep coming back, year after year. They are brilliant, beautiful creatures; watching them swim upstream together (Fig. 9.1) is an awesome wildlife experience. They fill our smokehouses, drying racks, freezers, and boat holds. Salmon have played a fundamental role in shaping the history and culture of Alaska.

Salmon populated southern Alaska's rivers and streams following the end of the last glaciation, about 10 000 years ago. Pacific Northwest aboriginal cultures flourished in large measure due to the abundance and food value of salmon; still, today, salmon are pre-eminent in the distinctive art forms and totemic crests of Alaska native people living in coastal areas and along the major rivers. European explorers and settlers were quick to realise the commercial value of salmon, and by the end of the 19th century the commercial salmon fishing industry was firmly established. The salmon economy and associated technology developed rapidly. By the early 20th century, much of the commercial catch was taken in large fish-traps, operated by the salmon canneries which dominated the industry. The canned salmon industry wielded enormous political clout, the salmon management capability of the US government was poorly funded and ineffective, and by 1940 salmon-runs already were severely over-fished. By then, even the salmon canneries pressed for better management. By the 1950s major salmon runs throughout Alaska's southern and south-west coasts were greatly depleted and in some cases decimated. Alaska entered the Union in 1959, amid demands by Alaskans for local management of the state's salmon fisheries. Indeed, the Alaska constitution, ratified by the voters in 1956, devotes an entire section to the management of natural resources. With the once-abundant salmon resource very much in mind, drafters of the constitution directed the legislature and executive branch to ensure 'Management of renewable resources on a sustained yield basis'. One of the first acts of the new legislature was to outlaw the salmon traps, forever.

Fig. 9.1   Salmon swimming below the Creek Street Bridge, Ketchikan, Alaska. (Photograph courtesy of Phil Doherty, © Alaska Department of Fish & Game.)

With a clear constitutional and legislative mandate, the newly-created Alaska Department of Fish and Game (ADF&G) developed a salmon management programme based on a deceptively simple premise: each year returning salmon must be allowed to escape interception and enter freshwater habitats in sufficient numbers to perpetuate the stocks at levels sufficient to provide for a sustainable harvest. Aided by abundant pristine habitat and generally favourable ocean conditions, Alaska's wild salmon stocks have made a phenomenal comeback (Fig. 9.2). Now, harvests are being sustained at levels far greater than had ever been thought possible. The salmon rearing in Alaska waters and returning to Alaska's streams support robust sport and commercial fishing industries, and remain the basis for a traditional subsistence economy and way of life in scores of Alaska native villages throughout the state.

In 2000, over 137 million salmon, about 322 000 t, were sold by commercial fishermen in Alaska. In 2000, fishermen were paid US$275 M for their catch. Many Alaskans, including Alaska native populations still depend heavily on subsistence-caught salmon for food and cultural purposes. The estimated replacement cost for salmon caught to meet subsistence needs in Alaska exceeds US$100 million annually. State and federal law give top priority to subsistence uses of salmon.

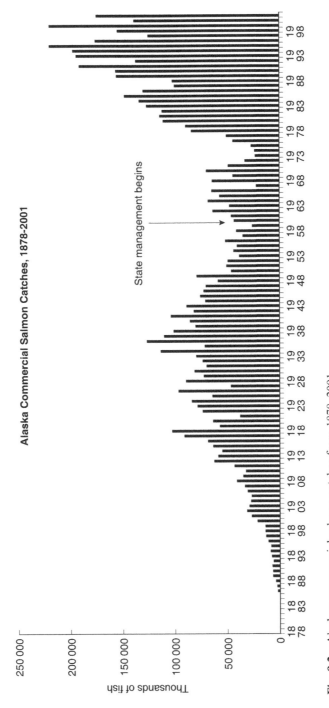

Fig. 9.2 Alaskan commercial salmon catches from 1878–2001.

9.8   Alaska and the MSC

The salmon restoration picture represented in Fig. 9.2 contrasts sharply with the reality along much of the rest of the Pacific Northwest coast, where wild salmon populations are declining, are at disastrously low levels, or have become extinct. Over-fishing has combined with water-use conflicts, development, dams, agriculture, and logging to create an environment adverse to salmon survival. More than just a natural disaster, the plight of salmon in the Pacific Northwest is the ominous back-drop for consideration of the future of salmon stocks and fisheries in Alaska. Every salmon manager in Alaska asks the question: how can we make sure the same thing that happened 'down south' does not happen here? Alaska is developing along with the rest of the world, our population is growing, and our natural resources are under enormous pressure from exploitation. We need look no farther than our own back-yard to understand the need for action to conserve the world's wild fish-stocks.

From the start of discussions with the MSC, many Alaskans saw a natural fit between the Alaska salmon fishery and the MSC certification programme. When the MSC programme was first discussed in Alaska, in 1996, the feeling among salmon managers at ADF&G was that Alaska had much to contribute to, and learn from, the global fishery-conservation debate. The MSC eco-label initiative provided a novel framework for that debate. Whereas fishery science and management, environmental activism, marketing, fishery economics and the pragmatics of business all eventually play a part in the Alaska salmon fisheries, the MSC brought these elements all into one sweeping, multi-national conversation. While this was new territory for salmon managers, it was easy to accept that salmon would play a prominent rôle in the discussion of ocean fisheries conservation. And as the world's largest salmon producer, Alaska clearly had a place at the table if we wished to be seated.

While the MSC's ultimate goal was fishery conservation, the prospect of eventually gaining rights to use the MSC eco-label looked also like a win-win-win for Alaska for more prosaic reasons: it appeared to be in the interest of the State of Alaska and ADF&G for our management programmes to gain international exposure, and for us to learn from that scrutiny; it seemed clear the salmon industry, which in 1996 was only beginning to feel market pressures from farmed salmon, would gain from the marketing benefit of the eco-label; and, there was clear evidence of a benefit to the growing number of consumers wishing to know more about the origins of their food.

While Alaska stakeholders were generally confident the state's commercial salmon fisheries would gain the right to use the MSC eco-label, if we applied, the prospect of Alaska becoming a part of the MSC eco-label programme nonetheless stirred sig-nificant debate. Some, including commercial fishermen, met the idea with scepticism. The MSC programme was new and untested, some thought it would result in added cost to fishermen, and others thought it might be an effort by environmentalists to meddle in the state's salmon management programmes. Ultimately, fishermen, processors, and fishery managers alike articulated the most serious concern. Early in the certification discussions they emphasised the need to avoid 'Balkanising' the state's salmon fishery, by focusing certification on only a few regions, thus setting up one region or gear group as being more sustainable than another. Allocation disputes among Alaska fishermen have long been substantially divisive, no one needed another issue to fight over.

As a consequence of these early discussions with the fishing industry, Alaska's overture to the MSC emphasised that our salmon-management policies and programmes apply to Alaska salmon generally, irrespective of the species or where they are caught. We described how Alaska's broad constitutional mandate for sustained-yield management compels this unified, conservation-based approach to the state's salmon fisheries. We pointed to statutory protections for salmon habitat that apply statewide, and showed that the state's sustainable salmon fisheries policy establishes guidelines and standards for salmon fisheries management throughout the state. Among its policy guidelines is the requirement that salmon in Alaska are managed to achieve necessary escapements; each year's return of salmon to Alaska is monitored in-season and management actions, including sport fishing restrictions, are taken to assure escapement needs are met. What is more, salmon from different regions are often inter-mingled at the processing facility. That the Alaska salmon fishery should be considered as a whole was the main qualification Alaska placed on its candidacy for MSC evaluation and eventual certification as a sustainable and well-managed commercial fishery.

9.9   The MSC assessment

Working with the certifier, Scientific Certification Systems Inc., Alaska proposed its commercial salmon fishery be considered as a test case that would combine both development and application of evaluation methods, standards, and performance criteria. This seemed necessary considering the fact that, at the time, no fisheries had been certified and specific evaluation procedures, criteria, and standards did not exist. The MSC agreed, and specified the test case would amount to a pre-evaluation of the fishery. When the test-case results were presented a year later they were sufficiently positive to persuade Alaska that its chances of receiving full accreditation were good. The state agreed in 1998 to proceed with formal certification, using the same evaluation team that had developed their methodology through the Alaska test case.

Government agencies occasionally must suffer internal audit, in which accountants from within the government poke and prod the agency's fiscal anatomy, and the salmon fishery evaluation conducted under the direction of Scientific Certification Systems (SCS) was not greatly different. In response to probing questions, ADF&G collected and copied annual management reports, fishery management plans, and research programme summaries. Site visits during the salmon season put evaluation team members in hip boots and spotter planes. Agency biometricians and their evaluation-team counterparts burned the midnight oil, huddled over spreadsheets. The MSC took pains to demonstrate they were 'hands-off' the evaluation process. That, and the fact the evaluation team was composed of already-busy people who resided outside Alaska, resulted in an evaluation process that left ADF&G largely on the sidelines, once information requests were fulfilled. The evaluation validated the MSC claim of an independent, credible, third-party review.

The evaluation report recommended the Alaska salmon fishery be certified and specified requirements to be met in order for ADF&G to maintain the certification. These requirements reflected the evaluation team's view that while the management and conduct of Alaska's commercial fisheries meet the MSC principles and criteria for a sustainable fishery; there is room for improvement. ADF&G considered the requirements for maintaining certification to be reasonable, and consistent with its own sustainable salmon fisheries policy for Alaska.

9.10   The salmon eco-label

Success of the MSC eco-label initiative depends on the premise that consumer demand will provide an incentive for the fishing industry and managers of the fishery to meet high-performance standards. So is this premise being fulfilled in the case of Alaska salmon? Certainly, there is reason to believe so. In the past few years, increased media attention has been given to fishery conservation issues worldwide. As the fishing industry, especially the salmon sector, has become increasingly competitive, several major retailers and restaurants have promoted the Alaska salmon eco-label and consumers overall are probably better informed about seafood choices. A year after the salmon certification award, dozens of seafood firms – including many of the world's largest – have completed chain of custody and other requirements needed to enable their use of the eco-label. In many cases retail or wholesale buyers, responding to real or perceived consumer demand, prompted the processors and distributors to gain the right to use the label.

Research conducted by the Alaska Seafood Marketing Institute indicates sustainability messages resonate favourably with salmon consumers: 37% indicate the knowledge that Alaska salmon are not endangered makes them more likely to purchase the product (Alaska Seafood Marketing Institute, 2000). On this basis, the salmon eco-label shows early promise of fulfilling the 'win-win-win' scenario presented above: the State of Alaska and its fishery managers have gained international recognition for maintaining a sound fishery-management regime; the industry has a new niche-marketing tool, and the consumer can choose to support a well-managed fishery when buying salmon.

The jury is still out on success of the overall MSC programme, but it seems unlikely consumers years from now will retreat from a belief in the value of sustainable, well-managed fisheries or an eco-label that helps them act on that belief. On the contrary, since the Earth Summit in Rio (1992) there has been a growing awareness and concern for sustainability of natural resources. Heightened public awareness is evident in the increasing number of emerging societies, organisations, scientific reports, and media productions throughout the world focusing on sustainable development and use of resources. In a recent opinion survey Alaskans strongly supported a series of sustainable fisheries and ocean protection management policies (Fairbanks *et al.*, 2001). Seafood firms will have their own pragmatic reasons to help ensure a sustainable product-line, and many will continue working to satisfy consumers who want to make environmentally responsible purchasing decisions. The MSC eco-label initiative remains a grand experiment, but where Alaska salmon are concerned it appears less experimental and more mainstream, by the day.

Case Study 3: The Thames Herring Drift-net Fishery

10

Paul Medley & Paul Nichols

10.1 Introduction

A pre-assessment of the fishery was conducted on 29–30 September 1997 by Société Générale de Surveillance (SGS) of the Netherlands. As a result of positive indications for successful certification from this initial phase, a main assessment was conducted during 8–10 March 1999. The pre-assessment and main assessment were both conducted in accordance with the requirements of SGS's fisheries management programme, which is the SGS group's programme for certification against the MSCs principles and criteria. The pre- and main assessments were initiated through the efforts and financial assistance of the Essex Estuaries Initiative of Colchester Borough Council, UK. The fishery is located in the general area of the Thames and Blackwater Estuaries lying immediately to the north east of London, UK.

10.2 Stock

Separate herring (*Clupea harengus* – Fig. 10.1) stocks spawn in a number of locations throughout the North Sea. The main three stocks spawn during the autumn and produce yields in the region of 250 000 t per year. In contrast, there are a number of much smaller spring-spawning stocks in the estuaries around the North Sea, of which the Thames herring stock is one, producing yields of a few hundred tonnes. Although adults from different stocks may mix during periods outside the spawning season, scientific evidence indicates that they are distinct stocks; that is, inter-stock spawning does not occur. Thames herring are characterised by being physically smaller and in possessing, on average, more vertebrae than herring of other North Sea stocks. Evidence from tagging studies and research surveys also supports Thames herring as being a distinct stock.

Fig. 10.1   Thames herring (*Clupea harengus*), UK. Photo: Fishbase.

The adult herring begin aggregating into schools to feed, and later to spawn, from November. This stock spawns exclusively within the Thames Estuary from late February to early May in waters to the west of 01°20′ east. After spawning, the fish disperse to feed during the summer. Spawning takes place on a number of shallow-water banks in the estuary. Dependent on water temperature, larvae hatch and metamorphose to acquire the adult form by July–August. They are recruited to the fishery by their third year, and begin spawning when they are three years old. Prior to recruitment, the herring commonly appear in sprat and whitebait catches.

10.3   Fishing activity

In theory, any applicant owning a registered vessel under 17 m could obtain a licence and enter the drift-net fishery. In practice, less than 10 drift-net vessels are usually active, due to the relatively low value of the herring. Vessels are generally less than 10 m length overall. Some small dories are also active and use the same drift-net gear. All participating vessels are day-boats, spending a few hours fishing at a time before landing, although when market prices increase some operators are induced to fish at night. The vast majority of vessels are based, and land their catch, at West Mersea in the Blackwater Estuary.

The drift-net gear comprises of nets approximately 400 m long set 4–8 m deep. Most vessels usually fish three nets but some set as many as six. The immersion time for the nets varies but is usually not more than two hours. Nets are brought in using a hydraulic hauler once the floats indicate that there are fish entangled in the nets.

Immediately to the south of the drift-net regulatory area an active pair-trawl fishery operates mainly between October and November. Thames herring constitutes a significant part of the income for local trawlers during this time. This area falls under the jurisdiction of the EU Common Fisheries Policy (CFP) and is managed accordingly. Although relations between the driftermen and pair-trawlermen are good, the trawlermen were not supportive of the MSC certification.

Herring go off-shore and disperse in deep water when not aggregated for spawning. No trawlers of other nationality are known to fish this herring stock in deep water and other catches are presumed to be inconsequential.

10.4   Post-harvest handling and distribution

Fish are landed un-iced to local cold storage facilities at West Mersea and East Mersea. The distribution system has three main routes:

(1)   supply to wholesale merchants, for national or European distribution and processing;
(2)   direct supply to local retail and catering outlets; and
(3)   supply to local processors (very occasionally) for production of rollmops, bloaters and kippers.

Merchants indicated that the landed buying price is highly influenced by the end market and product quality. Merchants believed that being a day fishery, the drift-net fishery has an advantage over multi-day herring fisheries, where quality is often poorer. Merchants interviewed appeared to be supportive of MSC certification. Chain-of-custody audit appeared to present no significant problems because the origin of all boxes of fish is clearly labelled and the distribution chain is short and easily monitored.

10.5   Fishery management

The regulated Thames Estuary drift-net area was established under a UK government regulation in response to concerns that trawlers were damaging the spawning grounds. The regulated area is defined as all waters within the area demarcated by 51°56' and 51°33' N and west of 01°19.1'.

The decline of the main North Sea herring stock resulted in increased catches of Thames herring peaking in 1972–3 with a record 606 t. Subsequent poor year classes and a decline in spawning-stock biomass led ultimately to the closure of the fishery in the winter of 1979–80. After an apparent rapid recovery with a strong 1980–81 recruitment, the fishery was re-opened. It has since been maintained through a strict TAC (total allowable catch) set annually by the UK government.

The UK Department for Environment, Food and Rural Affairs (DEFRA) is responsible for overall management. Normally, such management has to be consistent within the framework of appropriate EU Regulations relating to the CFP. However, the fishery is unusual in that it is officially recognised as not forming part of the EU fishery in ICES Divisions IVc–VIId. The derogation came about as a result of the *Sea Fisheries Regulation Act* 1966, which empowers the local management committee (see below) to manage inshore fisheries through bye-laws. Since the drift-net fishery does not fall under the auspices of the CFP system, this made the MSC assessment easier as decision-making is relatively more transparent.

The local management committee, the Kent and Essex Sea Fisheries Committee (KESFC), is responsible for local byelaws governing day-to-day management and enforcement for fisheries within the 6-mile inshore zone. Finally, a voluntary stakeholder organisation, the Herring Management Committee (HMC) provides a forum for community management.

10.6   Stock assessment

The stock assessment is used to set the total allowable catch (TAC) for the fishery. The TAC is set largely following scientific advice, which is given by the recognised scientific authority, the Centre for Environment, Fisheries and Aquaculture Science (CEFAS) and is set solely to conserve the stock. Although no technical document was available on the stock assessment, adequate information was obtained from other sources, such as discussion with the scientists responsible and non-technical documents.

There is an extensive time-series of data available for the assessment of this stock (1962–3 – 1998–9). Total landed catches are reported by fishermen. Landed catches are also frequently sampled mainly for data on length composition but also for age, sex, maturity and weight data. In addition to these fishery data, a fishery-independent survey (with mid-water trawl) is conducted by CEFAS each November to obtain fishery-independent indices of year class abundance. Effort is not directly recorded. By-catch and discards are not recorded. The data collected are suitable for virtual population analysis (VPA), a standard and widely used approach to stock assessment. This analysis estimates the state of the stock as assessed against biological reference points. Reference points are derived from the model and known biology of the stock and essentially define the point when a stock can be considered to be overfished as well as allowing clear definition of management aims for the stock. The system of reference points used in this fishery are based upon European and international standards and aim to maintain long term recruitment.

The stock assessment is used to set the TAC. Although the stock assessment takes full account of all past catches from this stock, the TAC does not. Thames herring caught outside the drift-net box by the mid-water pair trawlers off the Kent coast are not included in the TAC. Although the TAC is set correctly and enforced, the TAC alone cannot guarantee that fishing mortality will be limited to the target level since not all catches are included.

10.7   Stakeholder co-operation

The herring management committee (HMC) provides a forum for management of the TAC by setting voluntary individual quotas and also a mechanism for settling disputes. The HMC has no legal basis and operates voluntarily, with members drawn from fishermen, KESFC, DEFRA, and fish buyers. This voluntary committee has been successful largely due to the co-operation of the local fish buyers. The committee was set up to ensure that the TAC was fished gradually throughout the season after it was reduced in the late 1970s. The committee had become active again due to a desire by stakeholders to achieve MSC certification and also because of an appreciation that a more regular meeting of stakeholders is required than that afforded by DEFRA's annual pre-season management advice meeting. At the time of the assessment the new committee had met on one previous occasion. DEFRA convenes a meeting at the start of each season to present its management advice. This provides an opportunity for fishermen to discuss the condition of the fishery with scientists and regulators and to put their views across. The Essex Estuaries European Marine Site management includes participation of the fishing community and stakeholders in three advisory groups. Fishermen are able to appreciate the wider environmental management issues of the area and present their views on regulatory matters.

Given the complex nature of most fisheries, involving multi-species, multi-gear operations, it is important that certification is restricted to those fisheries where the fishers agree to behave responsibly (see FAO *Technical Guidelines for Responsible Fisheries: Fishing Operations*) and to afford complete and correct information for monitoring the fishery. This should be mandatory in support of the sustainable use of a particular stock.

10.8   Monitoring, control, surveillance and enforcement

As mentioned previously, the fishery falls completely within the area of competence of the KESFC. Some committee members are elected members of local councils and some are DEFRA appointees. KESFC has responsibility for formulating and enforcing management bye-laws within 6 nautical miles of shore, as prescribed under the *Sea Fisheries Act* 1966. A number of KESFC bye-laws are in force for Thames herring, including a drift-net vessel length limit (17 m), a minimum fish landing size (20 cm), a minimum mesh size of 54 mm for drift nets within the regulatory area, a minimum mesh size of 50 mm for trawl nets and a closed season when fish are actively spawning.

The UK government has direct responsibility for setting licence conditions, the TAC for the drift-net regulatory area and setting a minimum legal landing size. KESFC has a fast launch for enforcing local bye-laws and undertakes local patrols. The main reported problem, albeit small, was part-time fishermen who tend to be ignorant of, or ignore local bye-laws. As a result, enforcement efforts emphasise public awareness of regulatory measures in preference to prosecution wherever appropriate.

10.9   Environmental issues

An important support to MSC Principle 2 was the implementation of a protected area initiative. A large proportion of the regulated area is included in a proposed special area of conservation (SAC), which should be fully implemented by 2004. The SAC offers protection mainly in overseeing current management and statutory bodies to ensure that they carry out their environmental responsibilities. The area will be independently assessed to check its 'environmental health'. Should the herring fishery come into conflict with the conservation of the SAC, the relevant authorities are bound to act to reduce or eliminate any problems that arise.

Although no documentation exists on by-catch and discards, anecdotal evidence and direct observations were gathered. The by-catch was a very small proportion of the total catch, due to the highly selective nature of the fishing gear. A significant number of fish discarded from the drift-nets appeared to be alive on release. Fish of reasonable size, such as cod, are sometimes kept for personal consumption. None of the fish caught as by-catch appear to be endangered or threatened species, although this would be verified by a more thorough assessment of the by-catch. Seals are recognised by fishermen and DEFRA as a significant problem in the fishery, eating fish in the nets prior to hauling. There is no control of the seal population, although some research was being developed to reduce fish loss from the drift nets by discouraging seals taking fish from nets.

The River Thames and River Blackwater could be a source of a number of water-born pollutants. However, Thames water-quality has improved considerably over the past few decades, and there is no evidence of current water-quality problems. In addition, the SAC will help ensure that the relevant authorities continue to enforce water-quality regulations.

10.10   The MSC assessment

10.10.1   Fishery strengths

The environment within which the fishery operates will receive special protection from the proposed designation as a special area of conservation. This will help enforce current regulations on gear use and fishing activities, as well as protect against wider problems of pollution and environmental degradation, which may threaten the herring stock.

The stock assessment undertaken, considering the small size of the fishery, is extensive. The basic data, including the fishery independent survey, appear to be of good quality, although dependent on voluntary contribution from the fishermen.

The TAC is set with very clear objectives for conserving the spawning stock and is based securely on the scientific assessment. The TAC appears to be well enforced.

The fishing method is highly selective. The drift-net fishery has relatively little impact on other components of the local marine and estuarine ecological communities. There is no evidence that ecosystem balances and relationships have been changed.

The HM Committee represents an important, successful forum for co-management. This greatly improves the opportunities for long-term sustainable management.

The fishery itself provides two important employment opportunities. Firstly, the fishery provides fishermen with a chance to generate income when there are no other available fishing opportunities. Secondly, the fishery gives would-be fishermen a chance to start a career in fishing because of the relatively low entry costs.

10.10.2   *Fishery weaknesses*

The entire distributional range of Thames herring is not known. The TAC does not cover catches of the Thames Estuary stock outside the regulated area. These catches are counted against the southern North Sea herring TAC. These catches represent a significant proportion of the total catch of Thames herring.

There is no verification of data on landed catches. Recorded data on fishing effort, by-catch and discards is inadequate. Not all catches are recorded. Catches of pre-recruit fish are not accounted for. Given the fishery-independent survey and good recording of post-recruitment fish catches, this does not present too much of a problem in terms of stock assessment. Increasing impact on juvenile herring through, for example, increasing whitebait catches would result in poorer recruitment. Although the TAC could be adjusted, there would be no direct indication as to why there was poor recruitment, which would make appropriate management action difficult. There is a long-term need to improve catch recording and extend it to juveniles.

The potential exists for economic incentives to lead to non-sustainability of the fishery. The fishery is, in theory, open-access. Open access tends to undermine management controls and prevents the attainment of full economic and social benefits from the fishery. There was a concern that certification, by increasing the value of the fish, could exacerbate this problem. To prevent this, it was recommended that certification should be limited to those vessels that voluntarily submit to controls devised by the HMC.

While the stock assessment went beyond that which could reasonably be required considering the size and value of the fishery, the lack of technical scientific documentation on the stock assessment for more general review is a weakness. In general, such documentation should be maintained for all stock assessments to promote openness in the assessment process and allow improvements through constructive scientific criticism.

Finally, no account was taken of the socio-economic background when setting the TAC. Although in one sense this is a strength, since there is little pressure to increase the TAC for short-term gain, the lack of data and analysis on social and economic factors may lead to greater conflicts. Socio-economic issues are currently dealt with through the HMC. However, the lack of cultural, social and economic information could lead to unnecessary hardship for fishermen through imposed management controls and a lack of fair independent advice in solving disputes.

10.11 Conclusion

The Thames herring fishery was the smallest of the first fisheries to be assessed against the MSC principles and criteria. Expectations for data and information on the fishery for the certification process should be in accordance with the size and value of the fishery. For example, monitoring of fishing activities and stock assessment research is expensive and it seems unreasonable to expect the level of monitoring and research in a fishery worth, say, $10 000 000 to be the same as one worth $100 000. Without scaling ones expectations for the data and information available, certification would be increasingly limited to large-scale, valuable industrial fisheries. However, if the fishery size reflects the stock size, the fishery is still as likely to be subject to over-fishing as a larger more valuable stock. Simply scaling research and stock assessment rigour with the fishery size will not achieve MSC objectives.

In the case of Thames herring, the assessment team was fortunate. Besides the lack of documentation, which was easily corrected, the assessment that was undertaken was rigorous and of high quality. However, the assessment was not paid for by the participants in the fishery, therefore, this still raises the question as to what might reasonably be expected in the way of monitoring and stock assessment research given the size and value of a fishery of this size. The potential advantage that small-scale fisheries have over large-scale fisheries is the ease with which co-operation can be developed. Greater involvement in the management process by the fishing community not only decreases the need for and cost of compliance enforcement, but can also greatly improve the timeliness, reliability and quality of data and other information on the fishery. For the Thames herring fishery, there was a clear improvement in co-operation and community management. For stock assessment purposes, this could lead to adequate monitoring data. This, when coupled with slightly greater precaution when setting the TAC, should be adequate for ensuring that the fishery is sustained, even if the annual stock assessment survey is no longer conducted. Determining the appropriate level of research, monitoring, assessment and precaution for the fishery in question remains a subjective judgement, to be made by the assessment team.

The MSC certification process has had a positive impact on this fishery. Unlike most other management initiatives, MSC certification is requested by stakeholders, so any resulting improvements to management are more likely to succeed. Furthermore, the process encourages on-going co-operation by identification of specific aims. Improvements in documentation can also improve communication within the fishery. All these benefits were apparent in this fishery. However, the MSC-certification process does have some disadvantages. While increased documentation can be useful for communication, it is also costly and can lead to unnecessary bureaucracy if not kept in check. This is unlikely to be a big problem in the UK, where many of the necessary bureaucratic systems are already in place. For example, although not required for this fishery, formal stock assessment documents are produced for many other UK stocks by the appropriate scientific authority and therefore could easily be produced in this case. For many other countries, the sophistication of the documentation required for an MSC assessment may prove difficult and costly to achieve. Under such circumstances a more flexible approach may be required. Maintaining a consistent and valid certification scheme in the face of complex and diverse fisheries management regimes is being addressed by MSC, but will probably only be achieved through building precedents as fisheries are assessed.

Case Study 4: The New Zealand Hoki A: The Fishery

Edwin Aalders, Jo Akroyd & Trevor Ward

11

11.1   Introduction

The hoki (*Macruronus novaezelandiae*) fishery is New Zealand's largest fishery and one of its most valuable. Historically, the main fishery for hoki has operated from late June to late August in the west coast South Island (WCSI) area and from late June to mid September in the Cook Strait area. The fishery was developed in the 1970s by foreign fishing fleets, primarily from the USSR and Japan. In 1978 an exclusive economic zone (EEZ) was established by New Zealand, and the issuing of licenses to foreign vessels was restricted. Ultimately, in the late 1980s, foreign licensed vessels where phased out entirely from the hoki fishery.

In order to allow a more co-ordinated collaboration with all the different stakeholders, the hoki fishing industry in New Zealand founded the Hoki Fishery Management Company Ltd (HFMC). The objective of the HFMC is to enhance the management and economics of the hoki fishery through collaborative actions between the shareholders in research, management, organisation and promotion. In this role the HFMC took the initiative to apply for the MSC certification in early 2000. Following the assessment in October–November 2000 the MSC certificate was awarded to the HFMC for the hoki fishery on 14 March 2001.

11.2   The New Zealand fishery-management system

The New Zealand (NZ) hoki fishery is managed under the NZ *Fisheries Act* 1996, which is administered by the Ministry of Fisheries. The main management tool is the quota management system (QMS). The QMS was introduced in 1986 to control the total commercial catch from all the main fish stocks found within the NZ 200 nautical mile EEZ. The QMS is designed to ensure sustainable use of the fisheries resources while encouraging economic efficiency in the industry.

Fig. 11.1 New Zealand hoki (*Macruronus novaezelandiae*). Photo: Hoki Fisheries Management Company, New Zealand.

The approach used by the QMS is to directly limit the total quantity of fish taken by the commercial fishing industry, i.e. it is an output control. Each year the NZ government decides what quantity of each quota species may be caught. This is the total allowable catch (TAC) – the amount that can be taken by both commercial and non-commercial fishers. The TACC is the total allowable commercial catch. In the case of hoki, where there is no recreational or customary fishery operating, the TAC is the same as the TACC. The TACC comprises individual transferable quotas (ITQ). Within the commercial catch limit, access to fishing is determined by ownership of quota. Quota is a right to harvest a particular species in a defined area. Quota can be traded, bought, sold or leased. If the TACC is reduced in any particular year, individual quota holdings are reduced proportionally. Fishers do not own the fish in the sea; rather, ITQs give them the right to harvest a defined amount of the named species in a given area.

The Minister of Fisheries' decision on what the TAC should be is based on information supplied by the Ministry of Fisheries (MFish) and other interest and stakeholder groups including the commercial fishing industry, recreational fisheries, Maori and conservation groups. Scientists provide biological data such as the size of the resource and its productivity. The concept of maximum sustainable yield (MSY) is used to establish safe fishing levels for the stock. This is the largest annual take that can be taken over time without reducing the stocks productive potential. MFish evaluates the various risks to the fish stocks from particular catch levels in the future. Sustainable management of the hoki fishery is underpinned by an ongoing research and stock assessment process. About NZ$4–5 million is currently spent on hoki research each year. Scientists contracted by both the NZ government and the Hoki Fishery Management Company meet each year over a four-month period to analyse data collected over the previous year and to prepare management advice for the annual review of total allowable catches and other management measures.

Several different sets of data—collected by both scientists and by fishing vessel crews—are used in age-structured models to estimate size and productivity of the hoki stock and to predict trends in the stocks. Data inputs include:

- trawler catch rate, recorded daily by fishing vessels;
- size structure of the population;
- age data;
- environmental data;
- trawl surveys – hoki grounds are surveyed by research trawlers to determine hoki abundance and collect biological data;
- juvenile surveys, to estimate the number of young fish that will recruit to the commercial fishery in future years;
- acoustic surveys – specialised echo-sounders are used to survey the hoki spawning grounds and estimate the hoki biomass present.

11.2.1 Management controls

Most commercial fisheries around the world are managed by input controls. The main disadvantage of input controls is that overfishing is not necessarily prevented as one input can usually be substituted by another, e.g. restrictions on the number of vessels may result in those fewer vessels becoming more efficient, e.g. by fishing more often or with larger nets and vessels.

Output control, the approach used in the NZ QMS is to directly limit the total quantity taken by the commercial fishery. Although this system has been successful there are three key issues that need to be addressed:

(1)   There are only limited data for stock assessment of many of the species.
(2)   By-catch of non-target species – the onus is on the quota holder to ensure that they hold sufficient quota of the non-target species to cover the catch.
(3)   Other environmental impacts are hard to manage – the quota holder has the responsibility to ensure that all legal and other requirements of the fishery in relation to environment protection are properly discharged during the fishing operations.

These issues are being addressed by the Ministry of Fisheries and the fishing industry by way of additional research, legislation that provides severe deterrents for exceeding catch, and changes in fishing practices.

Although the hoki fishery is primarily controlled by output control there are other government input controls that apply to the hoki fishery. These include a minimum mesh size for hoki trawl nets, to minimise the capture of juvenile hoki, and a restriction on the size of trawlers fishing in designated areas around the coast. The Hoki Fishery Management Company also has a number of management procedures and monitoring regimes intended to improve the management of the resource. They are usually implemented through voluntary codes of practice. These include, for example, a code of practice to reduce catch of small hoki in order to increase the overall yield from the fishery, an agreement on the proportion of TACC that can be taken from each sub-stock of hoki, and codes of practice to minimise the by-catch of seabirds and marine mammals. For further information on the hoki fishery, consult the HFMC website (*www.hokinz.com*), and to contact the fishery about progress on these and other HFMC initiatives.

11.3   MSC principles and current fishery-management practice

To allow for a better relationship between the MSC principles and criteria, fisheries management of single-resource species needs to be replaced by more broadly-based management of the ecosystems that support all marine species. Protection for the ecosystems that support fisheries is essential. The management of the hoki fishery must be recognised as a subset of managing the whole ecosystem.

11.3.1   Ecosystem impacts

The hoki fishery operates in a range of ecosystems, mainly in waters beyond 200 m depth and often fishing at depths of 800 m or more. There is little knowledge of the NZ mid-water ecosystems but it is assumed from studies of similar habitats elsewhere (see for example Blaber & Bulman, 1987; Bulman, *et al.* 2001) that the mid-water ecosystems and continental shelf and slope habitats are diverse and productive. Removing large biomass of hoki from these ecosystems may have an impact on a range of species that depend on the hoki for food (such as seabirds, sharks, seals) or that are associated with hoki (such as hoki prey). The nature and extent of this impact is difficult to determine, and would be very costly, but it cannot be assumed to be insignificant, or even acceptable, given the lack of knowledge.

Although some hoki fishing is conducted near the seabed, the extent to which trawls make contact with the seabed is not known because the fishery does not have procedures in place to identify which trawls near the seabed make direct contact with the substrate. Depending on the nature of the substrates, therefore, the fishery may also have an effect on the ecology of seabed communities. Trawling in the mid-water depths where there is no contact with the seabed is likely to be much less environmentally damaging and effects are more limited to the indirect effects of withdrawing biomass of the target species from the mid-water ecosystem. In both cases, however, the hoki fishery appears to have an unknown effect on the ecosystems. Evidence worldwide suggests that mid-water ecosystems occur over large scales of space and time, and many are highly interconnected in terms of the distribution of mid-water species (Ward & Blaber, 1994). However, features of the seabed in deep waters (such as the shelf break and seamounts) may have very sensitive and locally restricted species that are of high biodiversity value and easily affected by fishing (see Tegner & Dayton, 1999).

Recent improvements in the hoki fishery have increased the catch of hoki within specific size ranges, minimising the catch of smaller fish, and increasing the economic efficiency of the fishery by focusing on a consistent size of fish that can be processed optimally by factory ships. The consistent harvest of a narrow hoki size-range increases the likelihood that the uniform application of a TAC applied as a fishery-wide output control could lead to substantial reduction in genetic diversity in the population by restricting the number of fish that can survive to grow larger than the size at which most fish are caught. The hoki fishery does not have any specific refuges set aside, and although this may not be an issue at present, there is a risk of adversely affecting the genetic diversity of the hoki population. Recent moves to apply an area-specific catch limit to hoki will reduce this effect. As well as making the fishery more sustainable, this could act to alleviate fishing pressure on genetic diversity in the hoki populations.

By-catch

The hoki fishery catches seals (New Zealand fur seal), seabirds (albatrosses and petrels) and a range of non-target fish (such as sharks and ling) incidentally during fishing operations. This by-catch is both unwanted by the fishery and has created considerable concern for environmental groups. The seals are caught by vessels fishing near the surface (down to 200 m depths) when the trawls are being hauled back to the vessel. Seals attempt to steal fish from the nets, and are occasionally caught either in the net or in the wire rigging that controls the net.

The ecological effect of the catch of several hundred seals each year in the hoki fishery (although claimed to be of the order of a thousand seals each year by environmental groups) is unknown. The present population of these seals is greatly depressed from the original pre-sealing population, and it is possible that the additional mortality imposed through fishing activities has an impact on the recovery rate of the seal population. This may be a local or population-wide impact, but this is presently unknown. The population of New Zealand fur seal is not considered to be at risk (it is not classified as a threatened species), and although fluctuations in the rates of pupping have been recorded in monitoring studies conducted by the NZ Department of Conservation, the overall impact of the fishery on these seals is not clear but is unlikely to be of major ecological concern. Nonetheless, to allay fears that there may be a significant impact on these seals, the fishery has been required to make further attempts to assess the importance of this problem, by supporting studies of the seal populations, and is also considering devices to reduce the incidental catch of seals in the fishery. While the by-catch of seals in the hoki fishery may not be of major direct ecological concern, the impact of this mortality has to be considered in the context of the total mortality suffered by seals of this species and the extent to which the other sources of mortality can be controlled and reduced.

The hoki fishery also has a by-catch of seabirds that become caught in nets and wires, and occasionally they collide with the fishing vessels. The NZ government monitoring programme has recorded about 1000 seabirds as by-catch each year (predicted numbers from the ship-based observer programme). These birds are mainly albatrosses and petrels, and include some species that are listed by the IUCN as threatened (on the IUCN Red List). The fishery impacts on these vulnerable species may be ecologically very important even though the number of such birds killed may be very small. Some of these seabirds, such as the long-lived albatrosses, have been dramatically reduced in number by other fisheries (particularly long-line fishing) and possibly climatic changes, have a naturally low fecundity, and breed in only very restricted localities. This means that, for some of these species, the recovery of their populations may be seriously affected by the additional fishing-induced mortality of only a few adult birds each year.

The impact of the hoki fishery on the seabirds taken as by-catch is not well understood but for the most vulnerable and at-risk species, such as the Campbell albatross, any additional mortality imposed by the hoki fishery is an important issue to be resolved. The need for robust and statistically valid monitoring data is paramount for such species, and where mortalities are occurring, it becomes essential to determine if such mortalities are sufficient to have an ecological effect on the bird populations. The importance of this as an issue to be resolved in the fishery can only be determined on the basis of reliable monitoring data. The statistical design of such monitoring should be based on being able to detect the by-catch of a few individuals of these most highly threatened species. Where good data are lacking for the most crucial species, monitoring programmes may need to be individually designed and conducted to provide reliable data on their by-catch.

Behavioural changes

Fishing boats are often followed by seabirds, and by some species of marine mammals, because such vessels in the past have often been a good source of easy food as by-catch or vessel wastes are discarded. Many marine species can be shown to display a behaviour known as entrainment, where they follow fishing vessels (and others) in the hope of being able to scavenge offal and discarded fish. Marine mammals, such as seals and whales, may be able to detect the acoustic signature of specific vessels, the unique sound of a vessel as it deploys its fishing gear, and have learned to target those vessels where by-catch has been previously discarded. Entrainment has major risks for such species and the disposal of wastes from all fishing vessels is either banned or strongly discouraged. Where populations of species are rebuilding it may be tempting to consider this a form of subsidy to the recovering populations but, unfortunately, such subsidies create dependencies that are not ecologically sound. While such food sources can assist species, such as birds, in the short term, in the long term the source is unreliable and may create a dependency that reduces the natural feeding capacity of the species concerned, and ultimately can be counterproductive to the populations.

11.4   Challenges and experiences

The general challenge of the hoki fishery assessment was to process the large number of documents and information in a short period of time set aside for the assessment team. However in the particular case of the hoki assessment the team was also faced with the specific challenge of facilitating the rôles and expectations of the various stakeholders that had, and wanted to have, input in the assessment process. The size and the significance of the fishery for both the industry and the country are such that any certification decision could potentially result in complaints from any of the participating parties. Keeping a balance between the different interests and expectations is the ever-present challenge with which all auditors are faced. This is particularly true in programmes such as the MSC that are new and have little or no normative reference points on how well the programme and the system behind it operates.

11.4.1   *Facilitating rôles and expectations*

Stakeholders

Stakeholders are an important partner in any certification and in particular within the MSC programme. Involvement of the stakeholders enhances the transparency of the overall process leading towards the ultimate certification of the fishery. Particularly in an early stage of a programme such as MSC, considerable attention has to be given to the stakeholders by providing them with information about the process leading towards certification and the rôle they can play within that process. The two main stakeholder groups are the governmental agencies and the non-governmental organisations (NGO) who often have a strong involvement in the fishery. It is also not uncommon that they will have conflicting perceptions of issues of how a fishery should be managed or corrected. Assessment teams, therefore, not uncommonly are faced with the fact that many parties have expectations of the assessment team that cannot be met due to the limitations of the certification process.

Governmental agencies

Under QMS system and the NZ legislation the management of the hoki fishery is the ultimate responsibility of the Ministry of Fisheries (MFish). The MFish responsibility covers both the management of the fish stock as well as the environmental impacts resulting from the fishery. Due to this clear delegation of responsibility both governmental agencies as well as NGO had the general expectation that the HFMC application for MSC certification would mean that the MSC requirements would have to be managed by MFish. This is a common perception that often occurs elsewhere in other countries with similar certification programmes. However, this would only be true if the government agency itself had directly applied for the MSC certification for the fishery. Also, in the MSC certification, none of the results of the assessment team's work have any reflection on whether a government agency is adequately doing its work, except where it may not be delivering on its mandate as specified in the relevant legislation and operating arrangements. As during the hoki assessment in NZ, all government agencies are a source of information that can be used by the assessment team to provide evidence about how effective the HFMC is in its management of the fishery to meet the MSC requirements.

Non-governmental organisations (NGO)

NGO normally take the role of 'whistle blowers' about impacts of the fishery that are not necessarily acknowledged by the industry or by governmental agencies that control the fishery. Thereby, the impacts of by-catch removal, fishing gear in benthic environments, and impacts on specific animal populations are some of the issues that typically receive specific attention from these groups. Whereas government agencies generally feel that they are audited in the process of a MSC certification, the NGO generally expect from a certification that the rôle of the assessment team is to introduce specific measures to reduce any negative impacts from the fishery that are identified by these NGO.

Within the hoki fishery, a clear example of this latter position would be the use of seal excluder-devices to reduce the by-catch of NZ fur seals. However, although the excluder might be suitable and effective for the circumstances, an assessment team will not normally prescribe the use of such a specific device. The assessment team may require that the number of seals in the by-catch be reduced to acceptable levels, possibly to zero, but will not normally prescribe how such an outcome is to be achieved. Such a prescription would ignore the possibility of a more innovative solution being developed by the fishery or the managers, and may preclude other forms of responses, such as limiting fishing effort in regions near seal colonies, that might be able to achieve the same outcome for seal by-catch. An assessment team is required to audit against the requirements of the MSC and, therefore, any specific usage of equipment can only be enforced by the assessment team if these have been specifically defined by the standard.

Fishing industry

In order to meet the MSC requirements, the HFMC has embarked on a number of initiatives that closely involve the various stakeholders. The most important of these are the development of a management plan for the fishery and an ecological risk assessment. Both these intended initiatives will involve considerable stakeholder consultation and participation, in both the design of the processes to be followed and in providing input material and information.

The MSC requirements for the hoki fishery to develop a detailed fisheries management plan and to conduct a comprehensive ecological risk assessment (ERA) have both resulted in the acceleration of plans by MFish in similar areas. New Zealand has recently begun the planning and stakeholder consultation process for design of an environmental management strategy for NZ fisheries, which is expected to contain considerable elements of similarity to the ERA proposed by the assessment team for the hoki fishery. Similarly, the MSC assessment identified the lack of an easily accessible and well-integrated management plan for the hoki fishery that covered all matters of policy, planning and implementation as a weakness in the fishery. New Zealand is now embarking on a major national initiative to develop and implement management plans for their fisheries.

11.5 Dispute resolution

Following the hoki fishery certification in March 2001, the MSC was faced with its first formal complaint from one of the stakeholders. The complainant, the Royal Forest and Bird Society of New Zealand, argued that both the certification system of the MSC as well as the decision to certify the hoki fishery were incomplete and incorrect. The MSC then established a formal dispute resolution process to deal with the complaint (see Chapter 6). The process involved a high-level review of the evidence that was presented and considered by the certifier's assessment team, to determine if the decision to certify the hoki fishery was correct, and what, if any, changes could be made to the MSC processes in order to remedy any problems that were discerned. At the time of writing, the outcome from the MSC dispute review had not been finalised.

B: The WWF Perspective

Katherine Short

11.6 Introduction

The hoki fishery was the first fishery to be assessed under the MSC programme in New Zealand, hence, the hoki fishery assessment process formally introduced the Marine Stewardship Council to the NZ environmental non-governmental organisation (NGO) community. The very essence of the MSC is its stakeholder base and WWF provided initial preparatory information about the MSC to key stakeholders in New Zealand. The hoki fishery assessment was also the first global MSC assessment and eventual certification of a white-fish trawl fishery.

11.7 Environmental NGO and NZ fishing politics

The environment and conservation community in New Zealand is strong, cohesive, and has two lead organisations. These are the Environment and Conservation Organisations of New Zealand (ECO), a coalition of over 55 groups from all over New Zealand, and the Royal Forest and Bird Protection Society (RFBPS) with many local branch offices. Also, Greenpeace and WWF, the international NGO, have nationally-based representatives. Greenpeace maintains a modest presence in New Zealand, committing itself to a small number of campaigns currently focussing on genetically modified organisms, toxics, climate and oceans. The RFBPS and ECO maintain a broad watching-brief on environment, resource management and conservation issues, with ECO producing a triennial *Vote for the Environment* policy statement in the lead up to the national elections.

In the 1990s, WWF New Zealand focussed its activities on a very successful classroom programme teaching young New Zealanders about conservation but was not a major player in the wider NZ conservation movement. Today, WWF New Zealand's conservation programme retains a strong environmental-education component but also includes policy and advocacy work on WWF's global priorities – oceans and coasts, forests, freshwater, species, toxics and climate change. Specific areas of WWF NZ work in 2001–2 included campaigning on albatross, petrel and marine mammal by-catch in fisheries, and certification of fisheries and forests. The WWF has a well-established community grants fund, in partnership with the Tindall Foundation, and has also initiated a partnership with the New Zealand development aid programme to support projects in the South Pacific.

The political situation between the nationally-based environmental movement and the fishing industry in particular is deeply mistrustful and adversarial. Whilst, in principle, supporting the use of certification as a means to improve environmental performance in forestry and fisheries, the nationally-based NZ environmental movement is cautious about the details of the certification assessment processes. They take a similar philosophical standpoint to that of the Western Australian environmental NGOs (see Chapter 8C) – the fishery should not receive certification until certain corrective actions have been taken, and proving sustainability is highly problematic. Overcoming these concerns requires substantial commitment on behalf of all parties involved and to date, the record has not been exemplary in any respect. Of the international NGOs, Greenpeace New Zealand tends to focus it's oceans campaign offshore in the Southern Ocean working on anti-whaling and illegal fishing, whilst WWF New Zealand is carefully and pro-actively building a working relationship with the national fishing industry.

WWF and others involved with fishery management in New Zealand believe the Marine Stewardship Council offers a new way of working together on the critical issues facing New Zealanders and those seeking livelihoods derived from *kai moana*, the food of the sea. However, incumbent on all involved with the MSC in New Zealand is the need to take a fresh approach. Indeed, the Hoki Fishery Management Company (HFMC) recently presented their case for pursuing MSC certification, to the FAO *Reykjavik conference on responsible fisheries*. This fresh approach requires the adequate resourcing of processes, using published guidelines and persevering with genuine effort to overcome the historical challenges and practical obstacles, and embedding the process of culture change into every level of politics, management, and operation.

11.8   WWF's rôle and the MSC

It is useful in the first instance to describe the three clearly distinct rôles WWF has with respect to the Marine Stewardship Council. These are: supporter and promoter of the MSC; facilitator of stakeholder involvement, and, commentator and monitor of fishery assessments and of the implementation of certification requirements, as a stakeholder in the process.

11.8.1   Supporter and promoter of the MSC

WWF Australia began a project in 1999 to establish the MSC in the WWF Oceania region (Australia, New Zealand, New Caledonia, Solomon Islands, Fiji, Cook Islands and Papua New Guinea) and, as part of this, established a partnership with WWF New Zealand to promote the MSC there (see Chapter 8C). This involved annual participation in industry and ECO conferences and national visits to a variety of stakeholders. Additionally, over this period the HFMC worked with Frozen Fish International, the German-based Unilever-owned major purchaser of hoki, to pursue MSC pre- and full assessments. However, it was not until November 2000 that the full assessment for hoki was publicly announced and the formal MSC process began.

11.8.2   Facilitator of stakeholder involvement

WWF gave initial advice to the certifier of the hoki fishery, SGS, through the SGS New Zealand office about the need to proceed cautiously with the MSC stakeholder process in New Zealand. Advice included recommending that SGS describe and publicise the process at the earliest point possible to the widest variety of non-government and government stakeholders. Emphasis was placed on the need to ensure transparency, demonstrate independent facilitation and the delivery of clear communications throughout the process. In conflict-based contexts, culture change is inevitably incremental but WWF believe that the critical components required for success with the MSC in New Zealand are a robust and clearly understood process and a willingness to address issues imaginatively. Suggestions were made to SGS about how to approach the stakeholder political situation to take these two ideas into account. The MSC is a courageous initiative, and the HFMC have been ambitious and forward thinking to have pursued certification. An ingredient lacking from New Zealand fisheries management throughout the 1990s was the freedom to be imaginative and seek alternative solutions.

An article in *Green Futures* magazine (September–October 2001) expresses this well:

'Samuel Taylor Coleridge once said imagination is the link between man and nature. We experience our world through our imagination and we should seek to continually exercise it. Taking this further, what the world looks like depends on *our* imagination ... Imagination doesn't always come easily, inspiration helps ... Whilst sustainability needs to be a creative process fuelled by inspiration, imagination and ideas, there are some practical tools that can help this process.'

This particular article describes the considerable boost that robust stakeholder processes can give to sustainability discussions. WWF is involved with a wide variety of stakeholder processes internationally in many areas of natural resource management. Experience clearly shows the correlation of clear communication, sound transparency and meaningful inclusiveness with greater, more robust natural resource management outcomes.

The final result of the assessment of the NZ hoki fishery was that it occurred, according to one participant, 'during the worst constellation of the planets'. All those involved, the certifier, the MSC, WWF, the national environment and conservation organisations, the industry and many from the government agencies were disappointed with the management of the process, and it failed to meet many of the expectations and requirements of the stakeholders and the MSC itself. The experience with the hoki assessment process in New Zealand reinforced the need, identified during the western rock lobster fishery assessment, for the MSC to develop robust stakeholder guidelines for certifiers to follow. These guidelines could include, for example, a controversy matrix asking questions such as: are there marine mammal interactions, does the fishery operate in a democratic society, are there environmental non-government organisations with campaigns focused on the fishing industry? If the matrix points to a significant number of potential controversies arising, then the certifier's proposal to the industry organisation seeking certification must factor in a sufficiently robust and comprehensive stakeholder process that allows meaningful discussion, consideration by the assessment team and, if necessary, appropriate negotiation of stakeholder concerns. The process must be clear and explicit to ensure that process issues do not prevent technical matters from being a focus, and do not distract from achieving improvements to the management of the fishery.

11.8.3   Commentator and monitor of fishery assessments

WWF's key issues of concern about the assessment process and the ecological impacts of the hoki fishery were expressed through submissions to the various stages of assessment, and covered five key areas:

(1)   overfishing of the eastern stock;
(2)   impacts on seabirds, some of which were listed as endangered, vulnerable and threatened;
(3)   impacts on New Zealand fur seals;
(4)   marine-ecosystem impacts, including trophic and benthic impacts, and
(5)   robust process.

At the conclusion of the fishery assessment the SGS assessment team issued ten corrective action requests (CAR) to address the points raised during their assessment. They required HFMC to produce corrective action plans to address these and required a six-month surveillance audit to check progress. One aspect of this process that became an issue from the environmental NGO point of view was that these CAR were deemed to be minor, i.e. the certification assessment team determined that the issues identified did not preclude certification. SGS later explained that the term 'minor' was a standard certification term they used to identify a corrective action that did not preclude certification (as opposed to 'major', which would prevent certification), and that subsequent surveillance and assessments would not use this term because of its more common meaning (something trivial) in English speaking countries. However, the ranking of over 500 fur-seal deaths and an unknown but potentially high number of seabird deaths per annum as minor was not well received by the environmental NGO community. This led to further mistrust of the process and created an obstacle to the effective marketing of hoki (see below).

Because of the controversy surrounding the hoki certification with respect to the interactions of the fishery with fur seals and seabirds, it is useful at this point to elaborate on the application of the MSC principles and criteria, particularly Principle 2. The ecosystem principle, Principle 2, is a major focus for WWF as it enables consideration of ecological and environmental issues previously thought by traditional fishery managers to be outside the immediate focus of fishery management. WWF believes the MSC principles and criteria provide an opportunity for an ecosystem-based management approach to be applied to fishery management more broadly (see US Department of Commerce, 1999, Ward *et al.*, 2002). Formally considering the ecological aspects of stock management, as part of Principle 1, is an example of one aspect of this approach.

The issue of whether a fishery is certifiable or whether it fails to meet the relevant benchmarks also needs to be considered in light of Principle 2. Marine ecological knowledge and information decreases the further one ventures out to sea and down in depth. However, this need not be an insurmountable obstacle for the MSC process. Certifiers can, and WWF believes should, require the fishery to apply modern ecological analysis to the management of the fishery, i.e. to the stock assessment, the harvest strategy, spatial allocation of effort, and fishery by-catch. Although certifiers are not able to recommend specific targets or reference points that management of a given stock must aim for, they can require that the fishery consider the trophic relationships, life-history characteristics, ecological impacts, and consequences of harvesting target stocks. They can expect the fishery to apply the most accurate and up-to-date best-practice solutions to minimise negative impacts on the ecosystem and target and non-target species. This is a unique strength and an important opportunity presented by the MSC approach. Over time, this body of ecological, specifically commercial fishery-related knowledge, should become increasingly valuable for future assessments, management and ongoing certification. It will enable the MSC and certifiers to determine more defined benchmarks and indicators and the exact point of pass and/or fail for a fishery on a given benchmark or indicator.

In 2003 SGS is due to release a report of the surveillance audit assessing the adequacy of the response to the CAR featured in the hoki corrective action plans. WWF has made a detailed submission to the surveillance audit team on these draft plans and will analyse the quality of the response against WWF's requirements. Having contributed to the establishment of the MSC, WWF is working within the framework that is provided by the certification of the hoki fishery to seek meaningful marine-conservation outcomes in the five areas listed above. Optimising the marine-conservation outcomes clearly requires substantial commitment and investment from all parties throughout the five-year life of the certification and WWF will work to foster this at each opportunity.

11.9   Dispute resolution

The Royal Forest and Bird Protection Society lodged a formal complaint with the certifier about the certification, 25 March 2001, requesting that certification be withdrawn:

'This complaint relates to our contention that the MSC Certification Report by SGS which concerns the New Zealand Hoki Fishery and dated 14 March 2001 fails adequately to interpret and to comply with the MSC Principles and Criteria and the requirements of certification. We ask that, in light of the failings of the report that the Certification be withdrawn.'

This detailed and thorough dispute is clear evidence of the dissatisfaction of environmental NGO stakeholders with the hoki assessment process that was run in New Zealand. Finding a negotiated resolution that produces an agreed outcome is a considerable challenge for both the certifier and the MSC itself. WWF considers this a critically important and necessary challenge if the MSC is to be credible globally and if the wider aims of culture change and reinvigorating fishery management in New Zealand with the freedom to be imaginative are to be achieved.

It is timely and useful that the MSC is developing new dispute procedures, however, at this time the hoki-RFBPS dispute remains unresolved and there are clear risks if this situation continues. There is also concern within the environmental-NGO community in New Zealand, and internationally, that the US $1000 lodgement fee for disputes will prevent disputes being raised in future and increase the risks to the MSC process and outcomes. These include risks to the credibility of the MSC, the confidence of the HFMC to maintain momentum internally, those marketing the products, and other fishery certifications if the environmental NGO in New Zealand promote their concerns internationally. Structures such as the MSC and its forestry counterpart, the Forest Stewardship Council (*www.fscoax.org*), rely on strong, broad and diverse stakeholder input. When difficulties arise, an issue in one sector may rebound into another, and so on. Preventing this and maintaining clear communication is paramount if these new independent certification mechanisms are to gain strength and meaningfully contribute to effective natural resource management.

11.10   Public perception

Whilst some of the fisheries management actions in New Zealand and the draft HFMC corrective action plans reduce the overfishing of the eastern stock and enable some mitigation of impacts on seabirds there is still a considerable amount to be done before real change 'in the water' is achieved. WWF believes the very public issues of the impacts of the fishery on NZ fur seals and seabirds are still not being meaningfully addressed, and this may become an impediment to the effective, proactive marketing of hoki in European and Australian markets that are sensitive to animal welfare issues. The absolute, scientific case may state that a given level of mortality caused by fisheries by-catch does not threaten the viability of some species of marine mammals or seabirds. However, often the information base leading to this is sketchy, the wider impacts unknown and the wise, precautionary approach is to aim to minimise mortality and seek further information at all times. Also, if the species in question is perceived to be attractive, there will be a problem with public acceptance. Where this is the case in European or westernised markets there is clearly a risk of high-profile public protests, such as 'bloodied mock-ups of creatures' being dumped on retailers' doorsteps, and the MSC being accused of 'misleading consumers'. WWF believes SGS did not require sufficiently robust action to mitigate seal deaths in the hoki fishery, relying on a risk assessment to determine any impacts and any subsequent action the Hoki Fishery Management Company might need to take arising from this risk assessment. Clearly, this is insufficient and risky and WWF has recommended to the MSC, SGS and those marketing MSC certified hoki, that this be addressed more fully in the corrective action plans.

The press release of the Environment and Conservation Organisations e-mailed at the time of certification (see below) reached a truly global audience of environmental NGO and management agencies:

'Hoki Fishery Certification Misleads Consumers – Condemned by NZ Environment Organisations Press Release by ECO at 14 Mar 2001 10:49

'The New Zealand hoki fishery is environmentally destructive and should not have been certified by the Marine Stewardship Council (MSC) says ECO, New Zealand's organisation of 65 organisations with a concern for the environment.

'The New Zealand hoki fishery drowns over 1000 fur seals each year, over 1000 seabirds of which 60% are albatross, die from the fishing,' says Cath Wallace marine co-ordinator for ECO. 'A multitude of animals are crushed by trawl nets when these scrape across the bottom.

'Hundreds of thousands of tonnes of hoki are extracted from the fishery each year with no environmental assessment of the impact of this extraction from the ecosystem or of trawling on the marine ecosystem. Few of the *Fisheries Act* environmental considerations are actually applied, and fisheries management proceeds in a state of considerable ignorance of environmental effects.

'There is a high risk (58–69%) that the Eastern hoki stock will collapse below 20% of the original unfished stock in each of the years to 2004 covered by the latest stock assessment. In fact it is more likely than not, to collapse.

'The legal target for this fishery is that stocks should be 33–35% of the unfished stock. When fish stocks drop below 20% of the unfished biomass they are commonly regarded as being at considerable risk. The stock risks collapse. There has been virtually no attempt to manage other impacts on the wider environment.

'The announcement that the hoki fishery is to be certified is a serious blow to the integrity of the MSC certification scheme. The MSC seems to have been desperate to get a portfolio of certified fish so that the scheme can get underway.

'With this certification, the MSC is sending a message that it will tolerate high levels of environmental damage and animal deaths. This is a disappointing loss of an opportunity to send reliable messages to consumers. Consumers will be offended by the MSC trying to give a green tick to a fishery with major environmental and animal welfare problems.

'Many of the problems we have identified were also noted in the certification report itself but the report then suggests that these are minor issues. We think that people worried about the environment and those worried about animal welfare will feel revolted at this attempt at greenwash in the face of the facts.

'ECO calls on the MSC to recognise that the assessment report's conclusions are not consistent with the facts and that this certification should be withdrawn. ENDS'

It demonstrated the reach of web-activism and the global reach of non-governmental networks. This must not be underestimated in future, especially with respect to potentially controversial certifications. Every effort needs to be made to construct robust and meaningful stakeholder processes to circumvent negative perceptions and potential consumer concern. Points raised throughout stakeholder processes need to be reflected in the assessment either by raising corrective actions, and ensuring consensus on whether these should be major or minor, or by justifying in the assessment report why actions are not raised.

11.11   Conclusion

The certification of the New Zealand Hoki fishery provides the MSC with a critical opportunity to demonstrate the inclusiveness of the MSC, its robust process and the ability of the MSC to respond to issues raised by stakeholders about MSC assessment processes. These demonstrations could enhance the credibility of the MSC. It has illustrated the consequences of some of the process gaps in the MSC guidelines and accentuated the need that these gaps, originally identified in the western rock lobster fishery certification, be rapidly filled. Whilst the western rock lobster fishery does export to the US, EU and east Asia, it is a highly seasonal fishery and consequently not as visible. However, the hoki fishery provides white-fish to the highly attuned, sensitive EU consumer through frozen white-fish products such as fish fillets and fish fingers. The issues facing the hoki fishery – the on-going by-catch of New Zealand fur seals and seabirds, and damage to sensitive ecosystems, and, to date, lack of a clear, published commitment to mitigate these – continues to prevent the pro-active marketing of the first MSC certified white-fish.

Finally, the five-year period of a certificate requires on-going engagement from stakeholders to really optimise the investment in time, effort and resources in being MSC certified. Stakeholders all have varying needs but the environmental NGO want to see their concerns about conservation and process issues meaningfully addressed. The industry needs to demonstrate internally that being certified creates a stronger, more secure, and profitable business. Buyers are aiming for greater market share, consumers want eco-friendly choices, and the MSC needs to grow, strengthen credibility, engage more fisheries in pursuing certification and achieve more depth in the stakeholder base. The 'poor constellation of the planets' description of the hoki assessment provides a clear opportunity for all participants in the hoki assessment, and in particular for the MSC and SGS, to embed the lessons learnt when proceeding with subsequent assessments.

Community-based Certification: A Route to Sustainable Fisheries

12

Julia Novy-Hildesley & Katherine Short

12.1 Introduction

During the early years of the Marine Stewardship Council (MSC), significant scepticism was expressed regarding the applicability of the MSC certification programme to community-based fisheries,[1] particularly those in developing countries. Developing country governments, philanthropic foundations, scientists and conservation practitioners raised this concern. In particular, critics contended that the costs associated with certification might prove beyond the means of poor communities, that certification would be problematic in developing states where fishery management is not based on extensive western scientific data, and the market incentive does not apply to many small-scale fisheries.

Fisheries play a very significant role in the economy and social life of developing countries, providing a vital source of protein and income to approximately 200 million people. In addition, the world's fish supply is highly dependent on supply from developing countries. Developing countries account for 50% of the international trade in fish, and community-based fishers (in both developing and industrialised countries) produce half of all fish and seafood consumed by people and represent 94% of all fishers (McGoodwin, 1994). As a result, what happens in community-based fisheries, and particularly in those in developing countries, greatly affects us all.

The United Nations estimates that at least 60% of the world's most valuable

[1] Community-based fisheries may be defined as relatively small-scale fisheries that are linked to a particular area where the community is based. In the literature, small-scale fisheries are defined as fisheries operating with a relatively low capital input and relatively high labour input (McGoodwin, 1990).

marine fish are either over fished or fished to the limit, yet worldwide demand for fish is projected to rise by 40% in the next decade. Considering that developing countries account for 50% of this international trade in fish, it is obvious that their involvement is critical to any global strategy to improve fishery management, such as the MSC programme. The challenge is ensuring that these fisheries are managed in a way that will guarantee continued economic, ecological and social benefits.

Along the world's coastlines, community-based fisheries provide most of the protein and jobs for neighbouring communities, yet are more threatened than industrialised fishers by coastal habitat destruction and pollution due to human activities on land. Many theorists and scientific researchers argue that artisanal fisheries have generally been more stable than those of the industrial sector because of the smaller scale and traditional fishing methods applied. However, uncontrolled participation and introduction of modern fishing methods from the developed world have contributed to declining catches in many parts of the world, often resulting in ecological decline and social disruption. The need to maximise short-term economic gain can encourage unsustainable use patterns, making fisheries vulnerable to over-exploitation and jeopardising the direct benefits derived from them. Moreover, fishing practices may damage the marine environment through uncontrolled use of destructive fishing gears and the extraction of marine species without consideration of biological, ecological and economic criteria for sustainability (CoBi, 2000). In addition to impacts on the productivity of the fishing sector, these recent trends in fishing have implications for larger ecosystems. Many community-based fisheries occur in regions known for their rich biological diversity. Thus, activities that harm the fishing sector can also harm these biologically important ecosystems. This in turn poses severe consequences for the future of the industry and these important ecosystems.

Certifying community-based fisheries could encourage the long-term sustainable development of such fisheries. While certification uses market incentives to entice fisheries of all sizes, the potential benefits to small-scale fisheries in developing countries can be particularly motivating. For community-based fishers, certification can result in access to new markets, an improved ability to compete with bigger, industrial fleets, and the potential to increase the fishing communities' standard of living.

12.2   WWF's approach to certification of community fisheries

Since the MSC was established in 1996, WWF has focused attention on the certification of community-scale fisheries, particularly those in developing coastal states. WWF believes the success of the MSC depends in part on its ability to overcome the obstacles associated with certifying fisheries of varying scales, including small-scale fisheries in developing countries. Recognising the need to create alternative models for resource extraction that incorporate the conservation of fishery resources, the security of fishers' livelihoods, and the protection of the marine environment, WWF initiated its Community Fisheries programme.

Community-based fisheries tend to occur in coastal, near-shore areas where the known greatest abundance and diversity of marine life often occurs and thus, their appropriate management is critical to support conservation efforts in many of the world's most biologically rich marine areas. WWF views the MSC as an important tool for protecting the marine web of life in these globally important areas, having witnessed the benefits of incentive-based conservation in a variety of sectors (Wilcox, 1994). As part of its *Living Planet* campaign, WWF collaborated with some of the world's leading scientists to identify the eco-regions most important to global bio-diversity – the *Global 200*. WWF has made a major institutional commitment to conserving these eco-regions in the coming years. Hence, supporting the application of the MSC programme in community-based fisheries is directly related to WWF's conservation policy in priority marine eco-regions of the *Global 200*.

12.3 WWF's proposed methodology for community-based certification[2]

Based on the MSC's certification process and WWF's experience working with fishing communities world-wide, WWF developed a methodology for certifying small-scale fisheries – community-based certification (CBC). CBC maximises the use of local knowledge and relies on partnerships with fishers and other stakeholders to assess the state of a fishery and to determine the relevance of certification to improve the management of the fishery. Initially, the methodology was developed to reflect WWF's practical experience pursuing certification with the fishing co-operatives in the Galapagos Islands of Ecuador. Since then, WWF has been engaged in certification projects in partnership with fishing associations, local and national non-governmental organisations, aid agencies, governments, and scientists around the world. The methodology for community-based certification reflects this breadth of experience, and remains a living approach and working document, incorporating lessons that emerge in new community-based certification efforts. The CBC methodology is divided into four linked stages, beginning with the selection of an appropriate site for community-based certification.

12.3.1 Stage one – is CBC potentially useful to the fishery?

In this first stage, the interested fishing association, NGO, or other entity is encouraged to screen a prospective fishery against a number of criteria to determine the feasibility of achieving certification in general terms. Success of the project will be affected by the availability of basic fishery information, such as:

---

[2] The term 'proposed methodology' is used to emphasise the evolutionary nature of the document and the variability across fisheries that may require different approaches to certification. It is also used to encourage other stakeholders to assist in developing and improving the methodology. For a complete copy of the methodology, visit: *www.panda.org/endangeredseas* and go to 'Community fisheries on the frontline'.

- fish stock status;
- gear types and fishing methods used;
- threats to sustainability;
- current market structure and prospects for 'green marketing';
- management controls and governance of the fishery;
- the community's interest in certification;
- funding; and
- local capacity to work on the project.

If stakeholders determine that the fishery is a good candidate for certification, the CBC methodology suggests working with the fishing community to develop a clear understanding of the MSC and the certification process. In some situations, a local fishing association may be the instigator of the project, in which case, the focus of WWF, local NGO, or other involved stakeholders is to provide technical support and information on certification to the community. In other situations, WWF, local NGO, or fish importers may be aware of the MSC, may identify a particular fishery as a potential candidate for certification, and may then work closely with local partners to introduce the certification concept to the particular fishery. This requires the collaboration of all stakeholders involved in the project, and an understanding of local customs and priorities. For example, private discussions with village chiefs or local leaders may be appropriate before larger community workshops or discussions are initiated. Through on-going dialogue, workshops and visits from other communities involved in certification projects, the community can come to understand and assess its interest in the MSC programme.

12.3.2   Stage two – informal assessment

The next stage in the CBC methodology involves an informal assessment of the status of the fishery with respect to MSC criteria and an exploration of economic opportunities provided by markets. Through this exercise, the community can evaluate whether pursuing certification is worth the time, collective community effort and financial commitment. The most critical aspect is the biological assessment which compiles existing information on the status of the fishery and reveals if further data gathering or analysis is needed for certification to proceed.[3] In some communities, this exercise may reveal that a comprehensive formal scientific stock assessment is not available for the fishery. If this is the only major shortfall, fishers and scientists may work together, with financial support from WWF, local governments, or NGO partners, to complete a formal stock assessment for the fishery, drawing on both Western science and traditional knowledge. This stock assessment can then be used in the MSC certification exercise.[4]

---

[3] Guidelines have been developed that translate the MSC standard into a methodology that can be adjusted and applied to most fisheries (*www.msc.org*).

[4] This is true for the lobster fishing Federation of Co-operatives in Baja California, Mexico and the Seri blue-crab fishermen in the Gulf of California, Mexico.

Other communities may use this exercise to create a 'road map' for their fishery, identifying the transformations that may be necessary to meet the sustainability criteria of the MSC. In the case of a blue crab fishery in the Philippines' western Visaya Sea, it was clear that a great deal of transformation was required before the fishery could meet the MSC standard. Nonetheless, WWF Philippines used the informal assessment to build relationships with local communities and to develop a strategy with them and the local government to support improved organisation of the fishing associations, to develop and implement management plans, and to fund enforcement of the fishery. While these efforts were initiated, WWF and the community decided to postpone plans for a formal MSC pre-assessment until the new measures were in place and proving effective (Jayme *et al.*, 2001). Hence, the CBC methodology is intended to help fisheries in all stages of development make their way through the MSC certification process. Clearly, some fisheries will require more work and greater transformation than others before proceeding to the final stage of the CBC methodology and undertaking a formal MSC assessment.

12.3.3 Stage three – market opportunities?

The market analysis component of this third stage in the methodology involves discussions with processors and buyers regarding the MSC programme and their interest in complying with chain-of-custody requirements necessary for the sale of an eco-labelled product. Some processors are not interested, because maintaining the chain of custody for the certified product may require them to process small batches of product separately (from the certified community-based fishery), thus losing efficiencies created by economies of scale when they mix products from different fisheries and process larger batches.

The market analysis also involves an exploration with select retailers committed to the MSC programme to determine their interest in purchasing products from the candidate fishery. (The MSC keeps a list of interested buyers.) Depending on the community, WWF, the MSC, hired consultants, or local NGO can play a rôle in facilitating these discussions with buyers. Furthermore, these discussions should help the community determine potential economic benefits of certification.

12.3.4 Stage four – formal assessment

The final step in the proposed methodology is the formal MSC certification assessment: a relatively brief pre-assessment followed by a comprehensive full assessment conducted by a qualified certification team. To initiate the certification process, the community selects a certifier. WWF or the MSC can suggest a list of approved certifiers, and the community may request a range of bids for the work.[5] Upon

[5] WWF has encouraged the MSC to support the approval of certifiers from developing countries. This should lower costs and allow developing country fisheries to access certifiers with local expertise and who are fluent in local languages. The MSC is actively promoting the approval of independent certifiers worldwide, by holding information meetings with potential certification companies in regions interested in fisheries certification.

completion of a formal contract between the approved certifier and the client (such as a federation of fishing co-operatives), the certifier carries out the preliminary assessment according to the MSC certification methodology.

WWF works with local partners and the MSC to ensure maximum participation and involvement of the community in the collection and provision of data and the incorporation of local knowledge in this assessment. In the pre-assessment exercise of the Seri people's blue-crab fishery, Gulf of California, for example, Seri governors shared their traditional ecological knowledge with the certifier and discussed their customary management practices which were then considered in the assessment exercise. Following completion of this relatively short visit by the certifier to the community, a pre-assessment report is submitted to the client. Depending on the results, a full assessment may be initiated immediately or postponed while the shortfalls identified during the pre-assessment are addressed. When the fishery is ready, a scientific team is assembled, possessing local expertise, and a full assessment is conducted. The same certifier who conducted the pre-assessment usually conducts this, but the clients may request quotes from other certifiers if they wish.

12.4   Who is involved in community-based certification?

WWF has worked with a range of partner organisations to facilitate community-based certification projects in over a dozen fisheries in Asia/Pacific, Africa, Latin America and Europe (see: *www.panda.org/endangeredseas*). These fisheries have become engaged with the MSC programme in a number of ways. In some cases, local WWF offices may propose certification to fishing communities with whom they have been working over the years. In others, an importer, local NGO, or aid agency may propose certification to a community and then contact WWF for technical and financial assistance. Elsewhere, a federation of fishing co-operatives or a fishing association may work independently with the MSC to become certified, or request WWF's involvement after learning about the community fisheries programme. MSC staff also engage directly with small-scale fisheries by promoting the MSC in rural communities and identifying partners who will support certification. WWF's community fisheries programme works closely with the MSC to provide technical assistance to interested fisheries, secure funding for projects, and enhance the MSC's ability to apply and adapt the MSC approach to these fisheries.

12.5   Motivation for certification in community-based fisheries

Community-based fisheries involved in the MSC programme have expressed a range of motivations for pursuing certification. Clearly, economic incentives are a major driver; however, the desire for community recognition, assurance of long-term sustainability, and political leverage also appear to be significant motivating factors (see Table 12.1). In north-eastern Brazil, a community is using the results of the

Table 12.1   Incentives for certification in community-based fisheries. (Adapted from WWF, 2000.)

Environmental incentives	Economic incentives
Desire for verification of long-term resource availability driven by catch declines; traditional management processes under threat; poor management by government agencies; threats to established rights to fish; resource allocation conflicts with other fishers, e.g. large trawlers.	Wider market access; increased leverage in markets; increased market security; improved prices; improved share in local or distant markets.

Political incentives	Community incentives
Kudos for local politicians; provision of methods-of-conflict resolution and bridge-building processes in community and local politics; provision of leverage in government processes; sense of stewardship strengthened and validated.	Increased community pride; building of confidence in resource management and future; validation of traditional, local knowledge; reinforcement of traditional systems.

preliminary analysis to lobby the Brazilian government for implementation of a national lobster management plan and has encouraged other communities to join the MSC programme by promoting certification at an annual regatta. In contrast, a Seri people's blue-crab fishery in the Gulf of California, Mexico, intends to use the MSC label to recognise the community's traditional management measures, in addition to differentiating its product from lower-quality crab entering the US marketplace.

A workshop to explore the relevance of the MSC programme to traditional, community-based fisheries held in Sydney, Australia, July 2000, (WWF, 2000) identified a range of incentives associated with certification. Workshop participants provided guidance for WWF's work in support of community fisheries conservation and contributed to the MSC standards-council process in which issues of community fisheries are considered. The group categorised the incentives as economic, political, environmental and community-oriented (Table 12.1). It was recognised that non-economic reasons for seeking certification seemed to have particular relevance to these fisheries. Table 12.1 illustrates how incentives in a small fishing community are often different from those recognised by industrial fisheries. Organisations involved in supporting community-based certification should seek to understand the incentives that drive each community in the context of the MSC programme.

12.6   Benefits associated with certification in community-based fisheries

Only a few community-based fisheries were MSC certified by 2003. These include the Thames herring fishery, the Burry Inlet cockle fishery, and a hand-line fishery for mackerel in the United Kingdom. In addition, nine community-based fisheries underwent formal MSC pre-assessment but had not completed full assessment. Some were working through the issues raised in the pre-assessment whilst others were contending with socio-economic or political obstacles. The nine include: the Seri people's blue-crab fishery in the Gulf of California, Mexico; the lobster fishery in Prainha do Canto Verde, Brazil; the Waddenzee cockle fishery in the Netherlands; the Southern Fishermen's Association mixed estuarine fishery in South Australia; a mixed hand-line fishery in Eritrea; a consortium of oyster fisheries in the United Kingdom; the Banco Chinchorro lobster fishery in Mexico, the Cayos Cochinos lobster fishery in Honduras, and the Baja California Pacific lobster fishery in Mexico. While it is too early to determine the extent of the benefits derived from actual community-based certifications, it has become clear that fisheries are benefiting in a number of other ways from their involvement in the certification process. For example, anecdotal evidence indicates that certified Burry Inlet cockles are being featured in London restaurants and the price for Thames herring has increased significantly since certification (MSC staff pers.comm., June 2001).

12.6.1   Management benefits

The process can advance the state of fishery knowledge by bringing leading scientists to a fishery. In the Meso-American reef, for example, internationally renowned scientists worked with local managers and fishers to develop new techniques for understanding lobster abundance, publishing their innovative findings in a peer-review journal (Phillips *et al.*, in press). In southern Australia, an information gap identified during the formal MSC pre-assessment spurred a new collaboration between fishing families and scientists to collect data on the impacts of fishing on the globally significant wetland where they fish. Similarly, the Seri people's blue-crab certification project gave impetus to bring together scientists from Sinaloa and Sonora, neighboring states in the Gulf of California, Mexico, to share data on the blue-crab resource. This compilation of data will facilitate the completion of a comprehensive stock assessment that can be used in the MSC full assessment. (CoBi project update, June 2000). In the Philippines, a certification project was the impetus for a genetic analysis to determine the extent of the blue-crab population fished by a group of communities seeking certification. The results will help them determine how large a stock assessment to undertake.

12.6.2   Market benefits

In addition to management-related benefits, a number of communities have found new market opportunities through the certification process. The Seri community and a Baja California Federation of lobster fishing co-operatives, for example, have developed new relationships with a tour boat company operating in their area. An early supporter of the MSC, Linblad Expeditions, is working with both communities to offer certified products to their passengers once the certifications are final (Linblad Expeditions Environmental Affairs Officer, pers. comm.). In addition, the Baja California Federation has developed new arrangements with European retailers and Californian restaurants. Royal Caribbean, the world's largest cruise company, has committed to purchasing certified products from the Meso-American reef and other community-based fisheries engaged in the MSC programme upon successful certification. In south Wales, UK, a community fishing for cockles achieved certification and their eco-labelled product is now featured in London restaurants.

12.6.3   Social benefits

Certification can also serve as a tool to bring people together to discuss the complexities of resource management. In the Philippines, WWF is working with fishing communities in the islands of the western Visaya sea, and has found that certification creates 'a venue to initiate a multi-sectoral drive for conservation where the government, business, fishing communities and individual consumers can collaborate' (Jayme *et al.*, 2001). In this fishery, a new dialogue was opened by certification that has since led to the strengthening of local fishing co-operatives, the development of regional management plans for blue crab, and the formal collaboration of WWF and the government to improve enforcement of no-trawl zones in the area (WWF Philippines, Project Report, June 2000).

Many communities also talk about a 'culture change' brought about by the certification process. Engaging in the MSC programme enables communities to think differently about their resources, approach obstacles from a different angle, and recognise and relate to the full set of stakeholders involved in the fishery, from fisher to consumer.

12.7   Obstacles to certification in community-based fisheries

There is also much truth in the criticisms raised by early sceptics. A number of factors may impede the progress of certification in community-based fisheries including the disjoint between traditional and modern knowledge systems, financing, and politics.

12.7.1   Knowledge systems

In some cases, particularly in fisheries governed by traditional ecological knowledge, applying the MSC principles and criteria is difficult. While the MSC standard is based on principles that are fundamental to the sustainability of any fishery, the format of the scoring guidelines currently in use does not make provision for fisheries that may be sustainable but are unable to prove as such. This is often because limited, if any, formalised data is available and traditional management systems quite different from those in industrialised fisheries are employed. This is often particularly true in community fisheries that fish or manage only a small portion of a larger stock or may fish in multi-species complex ecosystems. Because the MSC certification process requires the assessment of the entire biological stock of the species taken by the fishery, community-based fisheries often lack the necessary data to demonstrate sustainability. Indeed, this is an issue that goes beyond community fisheries. For example, the certification of tuna and other migratory, trans-boundary species is highly problematic, poor management in one region may preclude certification of the fishery in another even if the fishery in the other region is well-managed.

A number of small-scale fisheries have their own community-based or cultural systems in place that facilitate the sustainable management of resources, and it should be possible to assess such fisheries effectively against the MSC standard taking these factors into consideration. Such factors include traditional management systems, fishers' knowledge, informal control points for collecting data, and local customs or taboos such as the Polynesian *rahui* or closed area (Duncan Leadbitter, International Fisheries Director of the MSC, pers. comm. February, 2001). The MSC has recognised this issue and is working with WWF and others to support the development of scoring guidelines and performance indicators that will facilitate the evaluation of small-scale fisheries against the MSC principles and criteria. The MSC intends to support an expert-consultation to make recommendations on the feasibility of developing scoring guidelines and performance indicators that are appropriate to small-scale, traditionally managed fisheries. In addition, the consultation will provide advice to certifiers on mechanisms for interpreting information from such fisheries. For example, fishers' perceptions often hold important information but answers to fisheries assessment questions are often fine-tuned by the person giving the answer based on what they think the interviewer wants to hear or what will produce a favourable assessment. A solid grounding in local knowledge and ability to apply sensitive investigative techniques are required if such assessment methods are to be successful. The outcome of this effort to develop new scoring guidelines should be beneficial not just to developing country fisheries, but also to small-scale fisheries in developed countries and fisheries characterized by low data levels (Leadbitter, pers. comm.).

12.7.2   *Financing*

In addition to difficulties applying the MSC standard to traditional fisheries, financing certification is often beyond the means of small fishing communities. A number of costs are involved in certification, including fees for the certification company and its team of experts, costs associated with data gathering and analysis, and financing needed to make changes to fishing gear and other management transformations required by the certification criteria. MSC approval of certifiers in developing countries could significantly reduce certifier fees for community-based fisheries in less developed countries. However, there is still likely to be an important rôle for aid agencies, NGOs, philanthropic foundations and governments to continue to support the diversity of costs associated with certification for small-scale fisheries. Recently, the David and Lucile Packard Foundation supported the development of a new foundation to help finance MSC certifications worldwide, and undoubtedly, community-based fisheries will benefit from this.

12.7.3   *Securing the benefits*

Another issue of concern related to certification in community-based fisheries relates to the ability of communities to reap the economic benefits of certification. In some cases, community-based fisheries are committed to selling their products to one or two processors, from whom they receive fishing gear, fuel and financing for boat repairs. If these processors are not interested in certification, they may be unwilling to meet chain-of-custody requirements, disqualifying the product from use of the MSC eco-label. This means the community cannot derive economic benefits from the certification. In other cases, even if the processor fulfils chain-of-custody requirements, how can the community ensure that potential new profits derived from an elevated price are passed down through the chain of custody to the fishers? This may prove to be an issue for Seri people's blue-crab fishery where fishers are committed to selling all of their catch to one processor who has not yet committed to supporting the MSC process. In addition, many community fisheries have fairly limited, or only local markets. For example, roughly 80% of Pacific-island fisheries are subsistence based. Of the 20% that are market-oriented, only 10% are export-oriented (WWF, 2000). While interest in eco-labelling in Western markets is solid and improving, there is still uncertainty as to whether these mechanisms will have validity in non-western markets and a bearing on management of fisheries supplying those markets (WWF, 2000). While it appears that local companies that provide services to international tourists represent one potential proximate market, it remains to be seen whether developing country markets for eco-labelled products will grow.

12.7.4   *Politics*

Politics at all levels may impede certification. In some cases, certification exercises have been halted because a level of government has chosen not to disclose fisheries data to the certification team. In some developing countries the MSC is perceived as an initiative designed by industrialised nations and imposed on less-developed countries. While the MSC is a completely voluntary programme, some governments believe it interferes with the right of each country to manage its resources at will. In addition, some governments perceive the MSC as a potential trade barrier. If their fisheries fail to get certified, will they be excluded from important markets? And even if their fisheries are certified, what are the guaranteed benefits? Many governments prefer to validate their own management and not have their fisheries pay high costs to have outsiders assess their fisheries. Why should others be paid to assess the data that governments have paid to collect and analyse themselves?

12.8   Community-based certification in practice: a profile of the Seri people's *Jaiba* fishery, Mexico

The Seri community is one of the last remaining indigenous communities in Mexico. Like many coastal communities on the Gulf of California, the Seri's base an important part of their livelihood on small-scale fishing activities. This community operates a significant fishery for *jaiba* – a variety of blue crab.

Jaiba are caught in the Infiernillo Channel, in the central Gulf of California. Crabs are caught primarily with baited metal traps measuring roughly 1 cubic meter. Between 1500 and 2500 traps are set each year during a season of eight to nine months. The Seris' average annual catch is 625 t. The Seri community has exclusive fishing rights in the waters of the Infiernillo Channel. This concession limits the total fishing effort due to the small population size of the Seri Community (approximately 700 people). The Seri fishermen have extensive and detailed traditional ecological knowledge of their marine territory, a product of centuries of survival in the desert, with the ocean's renewable natural resources as the predominant food source. The Seris have an informal agreement with processors and government officials to close jaiba fishing during the breeding season. They also avoid the take of gravid females and have a recommended minimum catch size. While it is acknowledged that there is some lack of respect of minimum size limits, most Seris honour areas in which crabs are known to hibernate during cold months, as well as areas where females tend to concentrate.

In 1999, WWF established a partnership with *Comunidad y Biodiversidad* (CoBi), a local non-governmental organisation, to explore the potential for MSC certification with the Seri community in the northern Gulf of California. After a series of meetings with the Seri people's governor and other members of the traditional government, a formal collaboration agreement was reached in June 1999. CoBi worked to educate stakeholders about the certification process through a series of meetings with fishing co-operative leaders, fishermen and a local processor, as well as by producing and distributing a comic book about certification featuring a local folk hero. In addition, fishers became engaged in the monitoring of the fishery. CoBi staff gave a one-day course on jaiba fishery biology and monitoring techniques. During the course, the Seri elders' council shared their traditional knowledge of the jaiba fishery and promoted the engagement of young Seri's in new monitoring techniques. A grant was secured from the Arizona Sonora Desert Museum to support the salaries of four graduated Seri para-ecologists to be in charge of the jaiba monitoring activities during 2000–1.

Having laid the groundwork for community involvement in fishery monitoring, and achieved a basic understanding of the certification process, the community decided to proceed with MSC pre-assessment in March 2000. The stakeholders are interested in using certification to differentiate Mexican crab from lower-quality crab from other countries competing together in the US market. They also hope to market some certified crab locally to Linblad Expeditions, an interested tour operator.

Scientific Certification Systems, an MSC accredited certifier, completed a pre-assessment report in October 2000. The general conclusion was that the jaiba fishery has strong potential to continue to the full assessment phase once issues requiring special attention are addressed. Three primary shortcomings were identified:

(1) the lack of information on the biological status of the resource (a requirement of MSC Principle 1);
(2) the unknown impact of ghost fishing on the marine environment and the jaiba population (a requirement of MSC Principle 2);
(3) the lack of participation from a broad range of stakeholders in management decisions, including representatives from state governments that share the resource, and tribal and municipal governments (a requirement of MSC Principle 3).

To address the stock assessment issue, genetic work was commissioned to help identify whether or not the jaiba fished by the Seris is a sub-species of a broader meta-population. If this were the case, a local stock assessment would be sufficient to meet requirements of MSC full assessment. If not, jaiba data for all of the Gulf of California would need to be pooled together and analysed. The claws of 20 male jaiba per locality throughout the coast of Sonora are subject to allozyme and DNA analysis to check for discrete genetic sub-populations. If results of the genetic analysis indicate that the local jaiba is not biologically distinct from the larger population, scientists from Sinaloa and Sonora (the two most important jaiba producing states in Mexico) will jointly develop a stock assessment and combine their data.

Additional issues that need to be resolved include developing mechanisms for greater participation in management of the resource, and an analysis of the impact of ghost fishing. Forty per cent of traps are lost during normal fishing operations. The damage of these traps to the ocean bottom and to the extensive and highly productive seagrass meadows requires investigation and perhaps a change to more biodegradeable materials. CoBi has engaged a post-graduate student to analyse impacts of ghost fishing and is also working with both traditional and formal government leaders to discuss greater participation in the management of the fishery. The recent production of a jaiba video that features a discussion of certification will contribute to the educational process and encourage collaboration. As these issues are resolved, the Seris will prepare for MSC full assessment, and the MSC certifier will establish a certification team composed of internationally respected scientists and local experts.

12.9   The next steps for community-based certification

WWF acknowledges that community based certification (CBC) is a challenging objective. Nonetheless, the positive vision and cultural change it inspires motivates a dedication to the process. The Southern Fishermen's Association in South Australia is committed to achieving certification and is constructively addressing local political obstacles, declining environmental quality, and the short-term demand for subsistence fishing. WWF is seeking and developing partnerships with coastal management and technical institutions to refine and expand the CBC programme, to collate lessons, and to test the methodology in additional fisheries. Those fisheries already engaged with the CBC programme continue to work with their local WWF offices, fishery management agencies, NGO and other stakeholders to overcome obstacles identified during the pre-assessments, seek new markets, and address political challenges.

12.10 Conclusions

Certification of community-based fisheries, particularly in developing countries, is a challenging undertaking. Nonetheless, certification and eco-labelling offer a unique and powerful tool for community-based fisheries. The certification process allows communities to develop a 'road-map' for the future of their fisheries. By assessing their fishery against the MSC standard, communities may identify shortfalls in current management and develop strategies to address them. The road-map provides a comprehensive vision and facilitates a long-term management perspective.

Perhaps most significantly, the certification process creates opportunities for communities. Beyond providing access to new markets, certification connects communities to scientists, facilitates new dialogue between fishing communities and their processors, buyers, and government leaders, and inspires new innovative partnerships. These partnerships may, in turn, lead to a new way of viewing resources and a wider commitment to a sustainable future for the resources and the communities involved. The future of ocean stewardship relies upon innovative and adaptive tools that encourage positive change in fisheries of all scales. The MSC programme is an avenue to foster sustainability in fisheries that provide food security for millions and are located in some of the world's most biologically rich areas.

Is Eco-labelling Working?
A: An Overview

Volker Kuntzsch

13

13.1 Introduction

The introduction of the eco-labelling concept in marine fisheries has provoked reactions ranging from enthusiastic acceptance to harsh rejection across all stakeholders in fisheries worldwide. The Marine Stewardship Council (MSC) in 1996 was, and still is, the first globally applicable system to verify fisheries management practices against a standard that is based on the FAO *Code of Conduct for Responsible Fisheries*. NGO and those fishing industry companies that are involved with fish on a global scale generally welcomed this initiative, whilst smaller companies and local organisations questioned the applicability of eco-labelling in fisheries. The founding fathers of the MSC, Unilever and WWF, had different motivations to start the process but a common goal: to safeguard the future supply of fish.

Worldwide catches of the commercially important groundfish species, e.g. cod, haddock, Alaska pollock and hake have been in decline since the mid 1980s. The collapse of the cod fishery off Newfoundland in the early 1990s reflects the problem fisheries around the world were facing. Scientific evidence, when available, was often neglected for political reasons and quotas were set higher than recommended. In some areas quotas were in line with scientific recommendations but controls were ineffective and eventual catches were much higher. As a result, the total catch of the 10 major groundfish species almost halved within 15 years. These trends in global catches gave rise to concerns over future supplies of fish. It was a business decision that led to Unilever's engagement in setting up the MSC. The future of a successful fish business had to be safeguarded! This challenge became part of Unilever's sustainability agenda and was in line with consumers' expectations. To underline its commitment, Unilever made the pledge to source all fish from sustainable sources by 2005.

The MSC was created as an independent, non-profit, non-governmental organisation and was to work for sustainable marine fisheries while promoting responsible, environmentally appropriate, socially beneficial and economically viable fishing practices. It was envisaged to promote, globally, sustainable fisheries through voluntary certification. The certifications were to be carried out by independent, accredited certifiers on the basis of principles and criteria for sustainable fisheries. The principles and criteria were based on the FAO *Code of Conduct for Responsible Fisheries* and refined for certification purposes by fisheries stakeholders during a consultation process covering eight workshops around the world. Products from certified fisheries would be marked with an on-pack logo enabling consumers to select fish products they knew came from well-managed fisheries. The overarching purpose was to create economic incentives for well-managed fishing by harnessing market forces and consumer power through eco-labelling.

In 1998 the MSC became fully independent from its founding fathers. The scepticism and criticism that confronted the MSC since the beginning was to some degree aimed at its founding fathers. However, most reactions indicated that there was a misunderstanding of the workings of the MSC. The MSC was not supposed to be an institution responsible for additional fishery management regulations, nor was it to be responsible for certifications itself. Furthermore, the MSC was to be open for all stakeholders and was thus the only alternative for encouraging better fishery management practices globally. Since fisheries in the wild faced the more critical problems and had to be addressed more urgently, aquaculture was not to be considered in the first instance.

13.2   Implementation

Provoking improvements in ineffective fishery management practices requires a critical mass of already certified fisheries. With the MSC being relatively unknown, it was necessary for the industry, e.g. Young's Bluecrest Seafoods, Unilever, Kailis & France Foods, and NGO, e.g. WWF, to introduce the concept on a global scale. The large-scale fish processing industry had the advantage of sourcing from a variety of countries around the world, giving them the power of choice and the possibility to request from their suppliers the participation in a certification process. However, as fish catches decline, choice is reduced or may be unavailable and consequently, companies are not necessarily in a position to pressure their suppliers into a certification process. Furthermore, the involvement of all stakeholders in the certification process adds complexity in areas where suppliers are eager to participate, but other stakeholders are not. From the industry's point of view, actions needed to be twofold:

(1)   to create a greater understanding within the industry for the problems we are facing in world fisheries; and,
(2)   to work with suppliers towards a certification of their fisheries management.

Although there was sufficient knowledge about trends in world catches of fish, concerns were mainly addressed at the supply of single species. With the introduction of further species for human consumption in specific countries the urgency for resolving supply issues for the original species was neglected. Sudden price increases for some commodities increased the awareness of supply shortages, but changes in product portfolios on the supply and/or demand side countered these fluctuations and resulted in a cooling-off of the market. Country- or company-specific product preferences usually led to a specialised, but limited, view of the supply scenario. Therefore, it was necessary to address the supply of fish from a wider perspective and demonstrate the interdependency between different species belonging to a family, e.g. Gadidae, the cod-like fish.

Under the circumstances of limited supply, fisheries could not easily be forced by customers into participating in the certification process. Preferably, fisheries should be convinced through their understanding of the underlying problems in global fisheries and their contribution to improving this situation. However, more importantly, fishers needed to understand the certification process and eventual benefits that might follow the certification. Early discussions with fishers showed that the MSC and customers promoting the certification process need to know exactly what they are talking about. Frequently asked questions by fishers included:

- What does it cost me?
- How much more do I get for my fish with the MSC logo?
- What happens to my certificate when natural mortality reduces the biomass?
- Will the MSC move the goalposts in time to come?
- Will we get additional regulations?
- What happens if I want the certification but other participants in the fishery do not?

13.3   Where are we now?

Judging from press coverage and personal communications, the general state of world fisheries is now well understood by most stakeholders. This understanding has not only been brought about by introducing the term 'sustainability' at meetings and conferences, but also by the critical developments in fisheries that were previously regarded as well managed or sufficiently available, e.g. cod in the NE Atlantic Ocean and North Sea. Increasingly, industry leaders promote sustainability in fisheries with politicians, suppliers and competitors.

The first MSC certifications have been awarded to fisheries of differing scales – from Thames herring to Alaska salmon and New Zealand hoki. These certifications have not only been critical for refining the certification process, but they set examples for other fisheries to follow. Every certification has proven that fisheries are subject to natural fluctuations and that no fishery-management system can be perfect. The certification process itself does work for a variety of reasons:

- it is performed by an independent certification company;
- it enables all stakeholders to participate and voice their concerns, opinions, constructive criticism;
- it highlights areas that need further investigation or research to improve on the *status quo*;

- it involves international experts to review a fishery's management system;
- it improves the transparency of a fishery's management system.

The first products bearing the MSC logo have appeared in retail outlets and restaurants. Marketing activities are still limited because of the relatively small volumes of certified fish that are available to date. More certifications will follow. A number of pre-assessments are underway and interest in participating in the MSC process has been voiced from a variety of countries. The MSC has gained a remarkable recognition within the fishing industry, amongst politicians and NGO in a relatively short time. It is apparent, however, that all certifications have been with fisheries that are small and/or well managed. This should increase the motivation for other fisheries to follow but the real challenges still lie ahead. The concept of eco-labelling is successful only when a critical mass of certified fisheries has been gathered and the eco-label is required by a considerable proportion of the market. This is not the case as yet. To date, only about 1% of world fisheries have been certified and there is only limited visibility of the MSC logo. The MSC process has gained recognition with fisheries that can be regarded as certifiable but it has not directly provoked many changes yet in fisheries that are mismanaged. Underlying problems in fishery management, such as overcapacities and subsidies, have not been addressed as a result of eco-labelling. However, as the demand side has been getting more involved with sustainability, increasing attention is directed at fishery management by governments in countries where fisheries have been either neglected or driven by political objectives. This again underlines the importance of including all stakeholders in the process. The MSC cannot be left to their own devices; participation of the industry is required to implement sustainability on a global basis. Above all, it is in the industry's best interest to combat the depletion of the resources that ensure their business.

13.4   The next steps

In order to improve further on the MSC process and to achieve longer term improvements in fishery management through eco-labelling, it will be necessary to reflect on the following:

- additional certifiers will be necessary to cope with increasing demand for certifications;
- learning experiences from certifications need to be gathered and made available to fisheries requiring improvements in their fishery's management;
- the objective that receiving the certificate is a reward for good fisheries management practices needs to be further nurtured;
- the general knowledge within the fishing industry about developments in world fisheries needs to be further extended;
- the MSC process needs to be as transparent as possible to build confidence and trust with potential certification candidates and with consumers.

Within the relatively short time since the introduction of eco-labelling in fisheries, considerable progress has been made with the concept gaining recognition with fishery stakeholders around the world. However, the abolition of malpractices in fishery management will only be possible once a critical mass of certifications have been performed and visibility of the eco-label in the market place has grown. Only then, can we measure the real success of eco-labelling in fisheries.

B: Is Eco-labelling working – for Marine Ecosystems, WWF and the MSC?

Katherine Short

13.5 Introduction

WWF entered the global MSC partnership with Unilever to:

- arrest the decline in global fish stocks;
- improve marine ecosystems where fishery resources are taken; and,
- raise consumer awareness of the need to demonstrate fisheries products come from healthy and well-managed sources.

With respect to the ongoing development and implementation of the MSC, WWF has five informal roles:

(1) to increase awareness and understanding of fisheries certification and the MSC as a management tool to improve fishery management;
(2) to provide technical comment on the full-assessment stage of MSC assessments;
(3) to monitor, critique and assist with the implementation of any requirements arising out of the certification of fisheries by commenting on the periodic surveillance audits;
(4) to promote the MSC as a choice consumers can make to encourage the production of seafood from healthy and well-managed fisheries; and,
(5) to participate as a partner with certified fisheries to develop innovative responses to the challenges that arise in certification assessments and surveillance audits.

WWF focuses its activities in four main fishery product groups – whitefish, tuna, salmon, shrimp and prawn.

13.6   Arresting the decline

It might appear from global media coverage and the proliferation of worldwide conferences and research related to marine ecosystems, ecosystem-based management, and fishery impacts on the marine environment etc, that those involved with managing fisheries are aware of the crisis in marine fisheries worldwide. However, the increase in awareness and activity in conferences and discussion has not necessarily translated into strengthening fishery management or in a reduction in the threat to global fisheries. A contemporary situation was illustrated in the political interference played out in the European Union as the European Commission struggled to modernise the ailing Common Fisheries Policy and restructure it to be more effective in protecting fish stocks. In addition, even in countries deemed to have robust management techniques, such as New Zealand, Iceland and Australia, the collapses of fish stocks (cod in Iceland in 2001), the impacts of fishing on seabeds (New Zealand and Australia), the declines in major quota species (orange roughy in New Zealand and Australia), and the impacts of fishing on threatened and endangered species (across the world) still occur. There are also still many major fisheries where there are few or inadequate controls on catch limits, poor enforcement of fishery rules, little if any consideration of the wider impacts of fishing on the marine ecosystem, and major problems with unwanted catch.

13.7   Improving marine ecosystems

WWF's global marine programme has two main areas of activity: ensuring healthy and well-managed fisheries, and establishing protected areas. These are clearly interrelated and one may be a component of the other. WWF's approach to integrating these two areas is known as ecosystem-based management (EBM) of marine capture fisheries and is governed by five key principles (from Ward *et al.*, 2002):

(1)   Maintaining the natural structure and function of ecosystems, including the biodiversity and productivity of natural systems and identified important species, is the focus for management.
(2)   Human use and values of ecosystems are central to establishing objectives for use and management of natural resources.
(3)   Ecosystems are dynamic; their attributes and boundaries are constantly changing and consequently, interactions with human uses also are dynamic.
(4)   Natural resources are best managed within a management system that is based on a shared vision and a set of objectives developed amongst stakeholders.
(5)   Successful management is adaptive and based on scientific knowledge, continual learning and embedded monitoring processes.

The MSC provides a framework and systematic mechanism through which WWF can communicate these principles and foster actions that are consistent with achieving EBM. However, many of the underpinning management regimes required to implement the EBM approach in fishery management are poorly developed or struggling to achieve even the first building block—effective stock management. WWF believes the MSC is creating an incentive for industry to foster the development of these underpinning management regimes. The MSC ensures, through the required transparency and stakeholder based approach that this is done in partnership with others who have an interest in a healthy future for marine ecosystems and their resources. In the first instance, the MSC requires the basics of effective fishery management to be in place and these are codified in the MSC's standard – *The Principles and Criteria for Sustainable Fishing*. Second, a fishery is only certified once a particular suite of benchmarks or scoring 'guideposts' that interpret this standard have been reached. Thirdly, it is critically important to maintain this level and usually to address the issues that may have been raised during the certification. At each point there is the opportunity and need to inject marine conservation thinking to relate a particular aspect of the fishery to the wider ecosystem and to be innovative. At each point in the certification assessment and implementation process, WWF and other community-based, technical, management and industry partners can work together to foster greater understanding and realisation of these principles.

The opportunity exists for partners to trial alternative management approaches, experiment with elimination or mitigation of non-target catch, understand the ecological relationships associated with the target species, and to minimise other effects on the marine environment, such as through phasing out demersal trawling in favour of mid-water trawling. WWF believes that the above strategies can achieve healthier marine ecosystems and longer-term marine conservation outcomes. If MSC certification leads to fishery management considering (and eventually implementing) a network of marine reserves that might be appropriate for their unique suite of characteristics (necessary in terms of habitat protection), then the MSC can be considered to be contributing positively to marine conservation.

A fishery is certified for five years, providing time to undergo cultural change and shift communication, planning, research, and assessment resources that are necessary to implement any certification requirements and address any weaknesses in management of the fishery. In addition, if the fishery genuinely wishes to maximise the opportunity of being certified it can also plan to gain the greatest marketing benefit by building a five-year marketing plan and potentially, if all is on-track the certification can be rolled over into the next five year period subject to a new assessment. At this time, because of limited precedent and experience with the MSC programme, certified fisheries require more explicit guidance on seizing these opportunities, and there are a number of excellent and innovative approaches developing between certified industry bodies and their local stakeholders that are starting to do this. WWF believes there are signs in certified fisheries already that the 'penny is beginning to drop'. Key industry stakeholders are beginning to understand both the opportunity of innovation as well as the challenge of transparent rigorous engagement with those who historically have been antagonists.

The MSC can demonstrate that it is contributing positively to fishery management and broader marine conservation if, during the five-year life of a fishery certification, there is an improvement on issues such as actual levels of fish catch towards more sustainable levels and improvements in species protection and ecosystem conservation. WWF will monitor and comment on the MSC certification of fisheries, seeking to incorporate appropriate practices and requirements into fisheries management so that the fishery can justifiably be certified to be well-managed. After the first five-year period of certifications in the major fishery product areas of salmon, whitefish, tuna and shrimp, WWF will be able to evaluate the results and review the successes.

The MSC is also working with industry partners; brand owners such as Unilever, Youngs Bluecrest and Sainsbury; retailers such as Wholefoods USA, Migros (Switzerland) and also WWF to promote the MSC in the consumer market place. This will gain an increasing momentum with increasing volumes of frequently consumed, familiar products from fisheries such as the large volume white-fish fisheries (e.g. certified New Zealand hoki), salmon (e.g. certified Alaskan salmon), shrimp and tuna (WWF believes, however, that shrimp and tuna fisheries both pose enormous technical, biological and scientific challenges in determining appropriate levels of take and fishing practices.). Some of this promotion is already beginning to translate into real market gains. Some are receiving a better price for the product and an increase in their share of the market. Fisheries from different stocks of the same species are seeking certification as they begin to feel the market pressure to become MSC certified.

13.8   Conclusions

WWF entered the global MSC partnership with Unilever to arrest the decline in global fish stocks, and to improve marine ecosystems where fishery resources are exploited. The raising of consumer awareness to prefer fishery products from healthy and well-managed fisheries, the actual certification process, the MSC logo and the potential for financial incentives for fisheries and marketing intermediaries are all mechanisms to 'attain the primary objective'.

Much of the activity so far and some of the assessment processes have not been optimal. However, the MSC has worked with certifiers in a range of fisheries and is seeking to improve procedures based on the lessons learnt in each circumstance. This variability in the quality and timeliness of the assessment process has resulted in calls of condemnation from various stakeholder organisations that are critical of a perceived systemic failure of the MSC. WWF has sought to verify and strengthen some points in these arguments and to assist with improving MSC assessment and certification processes at every opportunity. Also, WWF will remain engaged with the MSC assessment processes for the first five-year period of the certifications of the major certified fisheries in the product areas of specific interest to WWF. WWF believes that only in this constructive way can relationships and trust be established, and genuine attempts protected to make stocks and marine ecosystems healthier. The strength of the MSC is that it is an independent organisation supported by the fishing industry, environmental organisations and other stakeholders. The MSC process is constantly evolving as evidence of its own commitment to continuous improvement, demonstrated recently by implementation of a new structure for MSC governance.

WWF remains firmly convinced that the MSC provides the only global mechanism to identify systematically well-managed fisheries. WWF will remain supportive, engaged and vigilant to ensure that high levels of innovation, transparency and thorough monitoring and reporting of actions are maintained in certified fisheries. Only through collective, collaborative partnership-based implementation will a pathway evolve bringing together the best endeavours of those who do want a healthier future for marine ecosystems. Whether this is to ensure a livelihood for their children, to know a fishery is in a better state than when the process was begun, or to ensure that marine ecosystems continue to function and evolve is not relevant. These are all valid reasons for engaging and will make the MSC what it should be – a force for ensuring healthy and well-managed wild capture marine fisheries for today, and for tomorrow's generations.

Conclusions

Trevor Ward, Bruce Phillips & Chet Chaffee

14

14.1 Introduction

Eco-labelling of fisheries is rapidly growing, fuelled by the promise of marketing benefits, more sustainable fish stocks, and fewer environmental impacts. The Marine Stewardship Council's programme is at the forefront of global efforts to introduce eco-labelling more broadly than at present, and although there are other smaller schemes, some locally based, others industry based, and some adapted from related environmental management systems, the MSC remains as the global flagship of eco-labelling in fisheries.

In this chapter we summarise the extent of progress so far with eco-labelling in fisheries based on the MSC programme, look forward to the next decade of activities to determine in what direction eco-labelling should progress, and discuss specific improvement that might be needed. We consider these matters from the perspective of the various interests and conclude with our own perspective on the future of eco-labelling in fisheries.

14.2 The incentives

The main incentives for fisheries are the marketing benefits. These benefits may arise for a certified fishery in a number of forms:

- increased price for fish and consequently improved profits;
- increased access to otherwise difficult markets;
- security in existing markets where competition is increasing;
- wider acceptability of products by consumers.

Other benefits for a certified fishery that may be somewhat more incidental, include:

- setting the benchmark standard for local, national and global best practice for both the certified fishery and other fisheries that might be managed by the same fishery agency or system (the 'rub-off' effect);

- more stable and secure financial and management system arrangements (related to the sustainability of the fishery);
- increased security of resource allocation in situations where there are other fishing sectors (recreational) or other fisheries in the same area that may reduce access to the fish stock;
- community and professional pride and standing that re-affirms a well-managed fishery (this is particularly important for local communities and community-managed fisheries);
- potential tax reductions;
- reduced regulatory oversight and controls;
- media counterpoints to negative messages about impacts of fishing;
- better relations with environmental and conservation groups;
- improved product quality.

All of these aspects have been discussed for other eco-labelling programmes with both documented and anecdotal evidence of their occurrence. Many of these benefits are already accruing to certified fisheries.

For conservation stakeholders the main objective is to secure improvements in fisheries' environmental performance. The extent to which the MSC can contribute to that objective is still unclear; however, the assessment process at the very least offers stakeholders the opportunity to identify issues that, from their perspective, are key to fishery sustainability and need to be addressed.

Findings of an MSC assessment are increasing in importance, in both the public and private sectors, elevating the debate about the significance of environmental issues in managing fisheries. All claims, whether from the industry, conservation groups, or government management and research agencies are subject to an analysis of the available evidence and an assessment of the matter against the requirements of the MSC standard. This, in theory, provides all stakeholders with the opportunity to secure significant outcomes on environmental or stock issues in the fishery, given the wording and the practice so far in implementing the MSC principles and criteria.

14.3   The outcomes

While it is as yet early in the MSC programme, a number of fisheries have been certified and although no fishery has yet to complete its full 5-year term of certification, some important outcomes have already emerged. Several certified fisheries have been able to increase their market penetration and, reportedly, prices and profits. The increase in prices can be determined by analysis of government statistics but increased profitability in a fishery is harder to determine and depends on many factors, including external market conditions and fluctuations in catch conditions etc. Also, fishery operators are somewhat reluctant to advertise better prices and profits because they argue that it may stimulate other fisheries to seek certification and hence, reduce the market advantage of the eco-label for their own fishery products.

The MSC programme has set the benchmark for global best practice of sustainability in fisheries. Not so much as a result of its standards (principles and criteria), because the FAO and others have drafted similar documents, but because of its implementation of external assessments where all aspects of good management must be evaluated by knowledgeable experts rather than taken at face value. In response to the emergence of the MSC principles and criteria, the process of global certification, and the outcomes of the first major certifications, national fishery managers in Australia and New Zealand have implemented fishery-sustainability systems modelled on those of the MSC and its early fishery certifications. Australia's national legislation and partner agreements with state agencies are closely modelled on the MSC principles and criteria (Fletcher *et al.*, 2002) but have yet to incorporate the facets of external review and documentation. New Zealand is developing a national environmental-management strategy for their fisheries that is built on the recent progress in the certification of the New Zealand hoki fishery (requirements include an ecological risk assessment which under the MSC became a requirement instituted by expert external reviewers to prove the importance, or lack thereof, of environmental issues in the fishery). These responses make clear that the MSC pro-gramme would never be seen as a replacement for locally relevant national sustain-ability-management systems but rather, the MSC may provide an overarching framework and perhaps high standards of achievement that can spur national systems toward higher levels of sustainability.

For some certified fisheries, the MSC process has established a framework and set of procedures where stakeholders can participate in the fishery management process in a more meaningful way. This means that fishery management processes are becoming increasingly open and accountable, and attitudes and assumptions used in the fishery-management system can be compared to a range of normative standards from outside the fishery sector. In most countries, wild capture fisheries are con-sidered to be public resources, with management devolved to government agencies and the fishing industry on behalf of the public. In this situation, stakeholders bring to the fishery-management system a set of standards and backgrounds that may be broader than those that can be found within the fishery sector itself. This provides an important opportunity for fishery management to be properly anchored in the standards of the broader community rather than those of a purely exploitative sector. Exposure to external analysis and participation of stakeholders are key attributes of the MSC programme that can promote and maintain sustainability.

In contrast, some certified fisheries are only slowly becoming engaged in delivering changes, even those mandated by the certification system. There is a long lead time for many of the required changes, related in some fisheries to the extensive internal consultative network that is required to ensure that the whole fishery is well informed and advised about the nature and need for changes in the fishery. These matters include the recognition of the need for comprehensive analysis of the ecological risks from a fishery, and the need for changes to the scientific research support provided to a fishery in order to target specific gaps in data and knowledge.

The long lead times for implementing real changes in fisheries has meant that the potential for changes that result in substantive environmental improvements have yet to be realised. While timetables for change have been established, realistically, the results of these changes are only likely to become expressed in the certified fisheries and be documented by the end of the five-year certification period. This means that few if any real improved environmental outcomes have yet been achieved in the certified fisheries. The ultimate test of a fishery's commitment to change and to long-term sustainability will be when a full re-assessment of the fishery must be conducted to maintain MSC certification at the end of the five-year term.

In summary, there appear to be good financial and other outcomes that are rewarding fisheries that have been certified. If these benefits are real, the MSC programme and the number of certified fisheries should grow rapidly in the next decade. In terms of improvements for ecosystems and non-target species, the MSC programme is proceeding apace and is full of promise, but at this stage there are few demonstrable outcomes. Both the critics and supporters of the MSC programme agree that eco-labelling has promise and, provided the implementation issues can be overcome, there will be a good range of benefits to be achieved.

14.4   The issues

From the certification perspective, a key issue to be resolved for the future is the tension between cost of a certification and the quality of analysis and assessment that it contains. There is a high level of pressure from fishery interests to reduce the cost of certification but the quality of the MSC programme, and its future, depends entirely on the use in each fishery certification of sufficient team members with a high level of expertise and adequate time to conduct the assessments properly. The MSC programme is different to most other certification systems in that it includes assessing real outcomes, in addition to the processes used in the fishery. This means, for example, the assessment team has to determine if the stock level has been maintained at an appropriate level, as opposed to determining if the fishery has an appropriate process for determining stock level, a considerably easier task. Hence, high-level experts with the appropriate experience are needed on each MSC assessment team and they need to be properly resourced in order to be able to make a defendable assessment of the fishery.

A further issue for the future is the ongoing synthesis and analysis of precedents in the programme, to ensure that internal best practice is always kept up to date and available for certifiers so that they can ensure their own practices are current and appropriate. This is closely related to ensuring that the MSC standard is always kept up to date and is expressed in a way that provides certifiers with the guidance that ensures consistency between fishery assessments but also supports and clarifies the role of expert judgement in the assessment process. To achieve this, the MSC will need to look critically at its standards, certification methods, the process for accrediting MSC certifiers, including the procedures for maintenance of that accreditation, and to the development and distribution of up-to-date guidelines on best case practice.

One area of the MSC that has lagged behind other activities is the promotion of the MSC label in consumer markets to power the market incentive. While some might consider this to be a certified fishery's responsibility, it is also that of the MSC.

Building the MSC 'Brand' should become a higher priority in the future with more pro-active advertising of the logo, and identifying it with global best practice in fishery management.

The future of the MSC also needs to involve more transparent engagement with stakeholders at the fishery level. The stakeholder role is a feature that sets the MSC process apart from other certification systems, and it is crucial to ensure that stakeholders are fully engaged and empowered to contribute to the assessments in an effective way. The MSC has already begun this process by a change in its own internal governance and adoption of a stakeholder council. However, there is also a need for much stronger guidelines on how certifiers should engage with stakeholders in each fishery to ensure that this component of the programme is properly implemented in each assessment. Improving the processes associated with assessments are crucial for the future, because sloppy certifications will ensure the MSC loses its standing, and will be decried by many who would have formerly been supporters. To date, certifiers themselves have carried the main responsibility for ensuring consistent high-quality assessments but the MSC should take up this responsibility and provide much greater guidance or operational training that certifiers can use effectively in their business environment.

Also important is determining what constitutes a high quality assessment. While a more strict scientific approach may be needed in complex assessments, this has to be applied with care, because this could significantly drive up the cost of certifications and limit them ultimately to only the most affluent fisheries. The key issue is the proper interpretation and application of the MSC principles and criteria. Without a base-set of performance indicators and scoring guideposts, certifiers have borne the burden of interpretation, thus increasing the risk of lack of consistency between assessments and increasing the possibility of a *de facto* lowering of the benchmark. Each set of performance indicators and the guideposts that spell out what constitutes acceptable performance need to be underpinned by genuine scientific principles.

The main risks to the growth and spread of the MSC are, paradoxically, mainly from the MSC itself. The MSC is poorly prepared for the business challenges posed by the certification system it has created. It is overly locked into a governance system that requires multiple levels of agreement and decision making but without the structure, capacity, expertise or funds to deliver support for this. This alone stands to bring the MSC to a standstill as a result of decision paralysis.

As the MSC programme has unfolded, it has become very obvious that the fisheries where urgent improvements are needed are not just those that are affluent, or exporting to markets such as Europe that are sensitive to eco-labelling, or based on cultures of environmental stewardship. Therefore, a major challenge for the MSC is ensuring that its programme can be used effectively in Asia and in community-based fisheries, which together account for about 50% of the world's fish catch. This will need to go well beyond language translations of the MSC documents. The basics of the MSC programme will need to be carefully evaluated and expectations for data and codification of practices modified to meet the basic cultural differences in these regions and arguably, to emphasise different aspects of the MSC principles and criteria.

Finally, the MSC has restricted its focus to wild capture fisheries until recently.

This has become an important concern because many of its supporters in the retail and food service sectors also need a programme that can address aquaculture production. The MSC is now evaluating the prospects of expanding into aquaculture. Many would argue that the future for seafood lies with aquaculture making it increasingly important that an MSC style programme for aquaculture is developed and implemented. There are many well-recognised environmental issues with aquaculture, and an MSC aquaculture programme would greatly assist in addressing these sustainability issues. An equally important issue, however, is the need to avoid any potential for displacing the negative environmental aspects of seafood production out of wild capture fisheries and into aquaculture. Any 'dumping' of the environment issues across sectors would be a false outcome for the MSC and simply create the façade of an improving global environment in the seafood sector. To avoid this an MSC aquaculture programme with standards matching those of the MSC wild capture fisheries programme is required.

In closing, this book has been written in order to document the experiences so far of people who have been involved with the implementation of the main eco-labelling programme for fisheries. We want to emphasise again here that the area is rapidly changing and the MSC programme itself is constantly being improved. Readers are advised to check the MSC website (*www.msc.org*) for the latest versions of the MSC programme and for updates on certified fisheries and the relevant issues. We also encourage readers of this book to become more acquainted with all the various certification activities involving independent assessments of fisheries and other seafood-production systems by contacting the programmes directly and getting involved. Good ideas, worthwhile expertise, and innovative solutions are required by these programmes. Given the opportunities these programmes present for educating consumers and improving fishery management and the seafood industry, we hope that more people with experience will engage in discussions and research to improve all these programmes.

References

Alaska Seafood Marketing Institute (2000) *Industry Study of Alaska Salmon in the US Market.* US Economic Development Administration, Washington, D.C.

Blaber, S.J.M. & Bulman, C.M. (1987) Diets of fishes of the upper continental slope of eastern Tasmania: content, calorific values, dietary overlap and trophic relationships. *Marine Biology*, **95**, 345–56.

Botsford, L.W., Castilla, J.C. & Peterson, C.H. (1997) The management of fisheries and marine ecosystems. *Science*, **277**: 509–15.

Botsford, L.W. & Parma, A.M. (2002) Uncertainty in marine management. In: *Marine Conservation Biology* (eds E. Norse & L. Crowder). Islands Press, Washington, D.C.

Brown, R.S. (1991) A decade (1980–1990) of research and management of the western rock lobster (*Panulirus cygnus*) fishery of Western Australia. *Revista de Investigaciones Marinas*, **12**, 204–22.

Bulman, C., Althaus, F., He, X., Bax, N.J. & Williams, A. (2001) Diets and trophic guilds of demersal fishes of the south-eastern Australian shelf. *Marine and Freshwater Research*, **52**, 537–48.

Caddy, J.F. (1996) *A checklist for fisheries resource management issues seen from the perspective of the FAO Code of Conduct for Responsible Fisheries.* Fisheries Circular no. 917. FAO, Rome.

Caddy, J.F. & Mahon, R. (1995) *Reference points for fisheries management.* Fisheries Technical Paper no. 347. FAO, Rome.

Casio, J. (ed.) (1996) *The ISO 14000 Handbook.* ASQC Quality Press and CEE Information Services, Milwaukee, Wisconsin.

Clarke, B. (2002) *The Good Fish Guide.* Marine Conservation Society, Ross-on-Wye, UK.

CoBi (Comunidad y Biodiversidad) (2000) *Community-based sustainable fisheries in Baja California: a pre-investment analysis to start a fisheries certifications program.* ITESM, Campus Guaymas – Terminación Bahía de Bacochibampo Guaymas, Sonora, Mexico, September 2000. WWF Mexico.

Deere, C.L. (1999) *Ecolabelling and Sustainable Fisheries.* IUCN, Washington, D.C. and FAO, Rome.

Dept. of Commerce (1999) *A Report to Congress by the Ecosystem Principles Advisory Panel.* U.S. Congress. Washington, D.C.

Duchene, L. (2001) Chefs help stir the eco-debate. *Seafood Business*, **20**, 24–8.

Dunlop, J.N. (2000) The assessment of ecological sustainability in Australian fisheries. *Focus in Fisheries*, **7**, 8–9.

EPA (Environmental Protection Agency) (1993) *Status Report on the Use of Environmental Labels Worldwide.* US EPA 742-R–9–93–001. Environment Protection Agency, Washington, D.C.

EPA (Environmental Protection Agency) (1998) *Environmental Labeling Issues, Policies, and*

Practices Worldwide. US EPA 742–R–98–009. Environment Protection Agency, Washington, D.C.

Fairbanks, Maslin, Maulin & Associates (2001) *Analysis of Alaska Statewide Survey*. p 4 of memorandum available at Alaska Department of Fish and Game, Box 25526, Juneau, Alaska 99802, USA.

FAO (1996) *Precautionary approach to capture fisheries and species introductions*. Technical Guidelines for Responsible Fisheries no. 2. FAO, Rome.

FAO (1999) *World Fisheries and Aquaculture Report* (1999) FAO, Rome.

FAO (2000a) *Fisheries Statistics: FISHSTAT*.
http://apps.fao.org/page/collections?subset=fisheries

FAO (2000b) *World Fisheries and Aquaculture Report* (2000)
http://www.fao.org/docrep/003/x8002e/x8002e00.htm

Fletcher, W.J., Chesson, J., Fisher, M. *et al.*, (2002) *National ESD Reporting Framework for Australian Fisheries: the 'How to' Guide for Wild Capture Fisheries*. FRDC project 2000/145, Fisheries Research and Development Corporation, Canberra, Australia.

Glowka, L. (2002) *Towards a Certification System for Bioprospecting Activities*. Grundlagen der Wirtschaftspolitik no. 2, State Secretariate for Economic Affairs (Switzerland).
http://www.biodiv.org/doc/meetings/cop/cop-06/other/cop-06-ch-rpt-en.pdf.

Govt. WA (1991) *Fisheries Management Act* (1991) Western Australia.
http://www.austlii.edu.au/au/legis/cth/num_act/fma1991193/

Hall, N. & Chubb, C. (2001) The status of the western rock lobster, *Panulirus cygnus*, fishery and the effectiveness of management controls in increasing the egg production of the stock. *Australian Journal of Marine and Freshwater Research*, **52**, 1657–67.

Hedlund, S. (2001) Spiny Lobster. *Seafood Business*, September, 50–1.

Hilborn, R. & Walters, C.J. (1992) *Quantitative Fisheries Stock Assessment: Choice, Dynamics and Uncertainty*. Chapman & Hall, New York.

ISEAL (International Social and Environmental Accreditation and Labelling Alliance) (2001) *ISEAL Member Standard-Setting Review*. Public Background Document, Issue 1.
http://www.isealalliance.net

Jayme, K., Apostol, R., Sariego, I. & Mider, J. (2001) *Lessons from the field: Promoting community-based certification of the blue crab fishery of the Northeastern Guimaras Strait, Negros, Philippines*. WWF Philippines, Quezon City.

Larkin, P. (1977) An epitaph for the concept of maximum sustainable yield. *Transactions of the American Fisheries Society*, **106**, 1–11.

Ludwig, D., Hilborn, R. & Walters, C. (1993) Uncertainty, resource exploitation and conservation: lessons from history. *Science*, **260**, 17–36.

McGoodwin, J.R. (1990) *Crisis in the world's fisheries: people, problems, and policies*. Stanford University Press, Stanford, California.

McGoodwin, J.R. (1994) Nowadays, nobody has any respect: the demise of folk management in a rural Mexican fishery. In: *Folk management in the world's fisheries: lessons for modern fisheries management* (eds C.L. Dyer & J.R. McGoodwin), pp. 43–54. University Press of Colorado, Niwot, Colorado.

MSC (2001a) *http://www.msc.org*

MSC (2001b) *Fish4Thought*. MSC, London.

MSC (2002) *An Overview of the scoring procedures used within the MSC Certification process*.
http://www.msc.org

NMFS (National Marine Fisheries Service) (2000) *Fisheries of the United States* 2000.
http://www.st.nmfs.gov/st1/fus/

Peterman, R.M. (2002) Eco-certification: an incentive for dealing effectively with uncertainty, risk, and burden of proof in fisheries. *Bulletin of Marine Science*, **70**, 669–681.

Phillips, B.F., Gonzalez Cano, J. & Vega Velazquez, A. (in press) Sustainable management of community-based spiny lobster fisheries and the problems of metapopulations. *Fisheries Research*.

Pitcher, T.J. (1999) *RapFish – a rapid appraisal technique for fisheries and its application to the* Code of Conduct for Responsible Fisheries. Fisheries Circular no. 947. FAO, Rome.

Pizzico, B. (2002) Eco-labels need industry backing to succeed. *Seafood Business*, February, p. 23.

SCS (2000) *Public Summary for the MSC Certification of The Western Rock Lobster Fishery, Western Australia*. Scientific Certification Systems, Inc., Oakland, California.

Sutton, M. (1997) A new paradigm for managing marine fisheries in the next millennium. In: *Developing and Sustaining World Fisheries Resources* (eds D.A. Hancock, D.C. Smith, A. Grant & J.P. Beumer), pp. 726–30. Second World Fisheries Congress Proceedings, CSIRO, Melbourne, Australia.

Tegner, M.J. & Dayton, P.K. (1999) Ecosystem effects of fishing. *Trends in Ecology and Evolution*, **14**, 261–62.

Viswanathan, K.K. & Gardiner, P. (2000) *Eco-labelling and developing country fisheries, threats and opportunities and the role of research*. (Abstract) Proceedings of Third World Fisheries Congress, Beijing, China.

Walters, C., Hall, N., Brown, R. & Chubb, C. (1993) A spatial model for the population dynamics and exploitation of the western Australian rock lobster, *Panulirus cygnus*. *Canadian Journal of Fisheries and Aquatic Sciences*, **50**, 1650–62.

Ward, T.J. & Blaber, S.J.M. (1994) Continental shelves and slopes. In: *Marine Biology* (eds L.S. Hammond & R.N. Synnot), pp. 333–44. Longman Cheshire, Melbourne, Australia.

Ward, T.J., Tarte, D., Hegerl, E.J. & Short, K. (2002) *Policy Proposals and Operational Guidance for Ecosystem-Based Management of Marine Capture Fisheries*. WWF Australia, Sydney.

Warren, B. & Haig-Brown, A. (2002) Eco-labeling rule change rankles pollock group. *Pacific Fishing*. February, p 20.

Watson, R. & Pauly, D. (2001) Systematic distortions in world fisheries catch trends. *Nature*, **414**, 536–38.

Weber, M.L. (2002a) *A Review of Global Ecolabelling Programs for Coffee, Forest Products, Marine Fisheries, and Marine Aquarium Organisms*. The David and Lucile Packard Foundation, Los Altos, California.

Weber, M.L. (2002b) *From Abundance to Scarcity: a History of U.S. Marine Fisheries Policy*. Island Press. Washington D.C.

Wessells, C.R. (2000) Ecolabelling and international seafood trade: the roles of certification costs and consumers willingness to pay. *Fisheries Economics Newsletter*, **50**, 44–9.

Wilcox, E. (1994) *Lessons from the Field: Marine Integrated Conservation and Development*. WWF, Washington, D.C.

Williams, S. (2002) *Using AHP and Expert Choice to Support the MSC Certification Process*. *http://www.msc.org*

WWF (2000) *Using the MSC in traditional, community-based fisheries and identifying candidate fisheries in the South Pacific*. Community Fisheries Workshop Report, WWF Australia, Sydney.

Index

Printed and bound by CPI Group (UK) Ltd, Croydon, CR0 4YY

Printed and bound by CPI Group (UK) Ltd, Croydon, CR0 4YY

16/04/2025

14658833-0002